The VSEPR Model of Molecular Geometry

Ronald J. Gillespie
McMaster University

István Hargittai
Hungarian Academy of Sciences
and Eötvös University

Allyn and Bacon

Boston   London   Toronto   Sydney   Tokyo   Singapore

Library of Congress Cataloging-in-Publication Data

Gillespie, Ronald J. (Ronald James)
   The VSEPR model of molecular geometry / Ronald J. Gillespie,
Istvàn Hargittai.
      p.   cm.
   Includes bibliographical references.
   ISBN 0-205-12369-4
   1. Molecular theory.   2. Molecular structure.   I. Hargittai,
Istvàn.   II. Title.
   QD461.G54   1991
   541.2'2—dc20   90-37928
                                                     CIP

Printed in the United States of America

10 9 8 7 6 5 4 3 2 1   95 94 93 92 91

Contents

3 The Valence-shell Electron-pair Repulsion Model   40

4 The Second-period Elements   62

5 The Main-group Elements of the Third and Subsequent Periods   98

6   The Transition Metals   160

7   The Quantum Mechanical Basis of the VSEPR Model   192

Subject Index   228

Formula Index   236

Preface

The valence-shell electron-pair repulsion (VSEPR) model of molecular geometry grew out of some ideas first proposed by N. V. Sidgwick and H. E. Powell in 1940 (N. V. Sidgwick and H. E. Powell, *Proc. Roy. Soc.*, A176, 153, 1940) and developed by R. S. Nyholm and one of the present authors in a paper published in 1957 (R. J. Gillespie and R. S. Nyholm, *Quart. Rev. Chem. Soc.*, 11, 339, 1957). This simple and widely applicable model continued to be developed in the following years, and a general account of the model and its applications was published by one of the present authors in 1972 (*Molecular Geometry*, Van Nostrand Reinhold, London, 1972). The ideas of the VSEPR model have become widely used in teaching molecular geometry in both general chemistry and inorganic chemistry courses, and brief accounts of the model appear in most general chemistry and inorganic chemistry textbooks. It has also proved to be a valuable tool for the discussion of molecular geometry at the postgraduate and research level.

Since 1972 a very large number of new structures to which the model can be applied have been determined. There has also been a considerable increase in our understanding of the model, particularly as a result of the work of R. F. W. Bader and his collaborators, which has provided a sound quantum mechanical basis for the model, which was previously lacking. Thus the purpose of this book is to give an up-to-date, comprehensive account of the model, its applications, and its theoretical basis.

The book will be useful to both undergraduate and graduate students in chemistry and also to their teachers, as well as to teachers of high-school and general chemistry courses. It will be of particular value to teachers and researchers who are interested in the physical basis of the model and in its relation to other methods of predicting and rationalizing molecular geometry. The first six chapters need no more than a general chemistry course as background and will be useful as a reference even for general chemistry students. Chapter 7 is a little more advanced and requires some knowledge of quantum mechanics for a full understanding.

Ronald J. Gillespie
Department of Chemistry
McMaster University
Hamilton, Ontario L8S 4M1
Canada

István Hargittai
Hungarian Academy of Sciences
    and Eötvös University
Budapest, PO Box 117, H-1431
Hungary

_1_

Molecular Geometry

This book is concerned with the geometry of molecules and with the interpretation and prediction of molecular geometry using the valence-shell electron-pair repulsion (VSEPR) model. In this chapter we review some basic ideas and concepts concerning the geometry of molecules. In the following chapter we briefly discuss the more important methods by means of which the geometry of a molecule may be determined. Then in the succeeding chapters we give a detailed discussion of the VSEPR model and use it to discuss the geometry of a wide variety of molecules. In the last chapter we consider the theoretical basis of the model and compare it with other models for rationalizing and predicting molecular geometry.

ATOMS, MOLECULES, AND BONDS

A molecule consists of a discrete group of two or more atoms held together in a definite geometrical arrangement. Whenever two or more atoms are held together sufficiently strongly to form a molecule, we say that there are chemical bonds between each atom and its close neighbors. The geometry of a molecule has a profound influence on its properties, and so ever since van't Hoff and le Bel proposed in 1874 that the bonds formed by a carbon atom have a tetrahedral arrangement the geometry of molecules has been of great interest. Chemists have for a long time represented a bond by a single line. We draw structures for molecules that indicate how it is believed that the atoms are connected together by bonds. We will use the term *structure* in this sense to indicate simply the connectivity of the atoms, while by *geometry* we mean the actual three-dimensional arrangement of the atoms.

Each atom in a molecule consists of a positively charged nucleus surrounded by a number of negatively charged electrons. Thus there are two important and closely related questions that we can ask about a molecule:

1

1. What are the relative positions of the nuclei in space? In other words, what is the geometry of the molecule? The geometry of a molecule is usually described in terms of the distances between the atomic nuclei that are bonded together (that is, the *bond lengths*), the angles between the bonds formed by each atom (that is, the *bond angles*), and the angles between the bonds on adjacent atoms (that is, the *torsional angles*).

2. How are the electrons arranged? Because electrons are in constant motion and because their paths cannot be precisely defined, the arrangement of the electrons in a molecule is described in terms of an *electron density distribution*. The electrons in an atom are arranged in successive shells surrounding the nucleus. The nucleus and the inner shells of electrons usually remain unchanged in molecule formation, and it is only the outer shell, known as the *valence shell*, that is modified. Thus the atomic nucleus and the inner electron shells are considered to constitute the core of the atom so that a molecule is thought of as consisting of two or more positively charged atomic cores held together by the electrostatic attraction of a negatively charged electron density distribution derived from the valence-shell electrons of the constituent atoms. Thus the relationship between the arrangement of the electrons, that is, the electron density distribution, and the bonds that are imagined to hold atoms together in a molecule is of fundamental importance to the understanding of molecular geometry.

The arrangement of the nuclei in a molecule may be determined by several different experimental methods, the most important of which is X-ray diffraction by crystalline solids, as will be described in Chapter 2. The geometry of a molecule may also be found, at least in principle, by determining by quantum mechanical calculations the arrangement of the nuclei that has the minimum energy, as will be discussed in Chapters 2 and 7.

The electron density distribution of a molecule can be determined by the same quantum mechanical calculations as are used to find the energy and geometry of a molecule. But in practice it is only possible to carry out these calculations to a reasonable accuracy for small molecules consisting of light atoms.

The electron density distribution can also, at least in principle, be determined by X-ray diffraction studies on crystalline solids. X-rays are diffracted by the periodically varying electron density distribution in a crystal; but most of the electron density is concentrated in the atomic cores, and so, although it is relatively simple to determine the positions of the atomic cores and therefore of the nuclei, it is often very difficult to detect the small changes in the electron density distribution that occur on molecule formation.

LEWIS STRUCTURES

The arrangement of the electrons in an atom is usually described in terms of its electron configuration as deduced from atomic spectroscopy, ionization energies, and the periodic table, and also from quantum mechanics (Chapter 7). The electron configurations of the elements are given in Table 1.1.

TABLE 1.1 ELECTRON CONFIGURATIONS OF THE ELEMENTS

Atomic Number	Element	Electron Configuration	Atomic Number	Element	Electron Configuration
1	H	$1s^1$	53	I	$[Kr]\,4d^{10}5s^25p^5$
2	He	$1s^2$	54	Xe	$[Kr]\,4d^{10}5s^25p^6$
3	Li	$[He]\,2s^1$	55	Cs	$[Xe]\,6s^1$
4	Be	$[He]\,2s^2$	56	Ba	$[Xe]\,6s^2$
5	B	$[He]\,2s^22p^1$	57	La	$[Xe]\,5d^16s^2$
6	C	$[He]\,2s^22p^2$	58	Ce	$[Xe]\,4f^26s^2$
7	N	$[He]\,2s^22p^3$	59	Pr	$[Xe]\,4f^2 \quad 6s^2$
8	O	$[He]\,2s^22p^4$	60	Nd	$[Xe]\,4f^4 \quad 6s^2$
9	F	$[He]\,2s^22p^5$	61	Pm	$[Xe]\,4f^5 \quad 6s^2$
10	Ne	$[He]\,2s^22p^6$	62	Sm	$[Xe]\,4f^6 \quad 6s^2$
11	Na	$[Ne]\,3s^1$	63	Eu	$[Xe]\,4f^7 \quad 6s^2$
12	Mg	$[Ne]\,3s^2$	64	Gd	$[Xe]\,4f^75d^16s^2$
13	Al	$[Ne]\,3s^23p^1$	65	Tb	$[Xe]\,4f^9 \quad 6s^2$
14	Si	$[Ne]\,3s^23p^2$	66	Dy	$[Xe]\,4f^{10} \quad 6s^2$
15	P	$[Ne]\,3s^23p^3$	67	Ho	$[Xe]\,4f^{11} \quad 6s^2$
16	S	$[Ne]\,3s^23p^4$	68	Er	$[Xe]\,4f^{12} \quad 6s^2$
17	Cl	$[Ne]\,3s^23p^5$	69	Tm	$[Xe]\,4f^{13} \quad 6s^2$
18	Ar	$[Ne]\,3s^23p^6$	70	Yb	$[Xe]\,4f^{14} \quad 6s^2$
19	K	$[Ar]\,4s^1$	71	Lu	$[Xe]\,4f^{14}5d^16s^2$
20	Ca	$[Ar]\,4s^2$	72	Hf	$[Xe]\,4f^{14}5d^26s^2$
21	Sc	$[Ar]\,3d^14s^2$	73	Ta	$[Xe]\,4f^{14}5d^36s^2$
22	Ti	$[Ar]\,3d^24s^2$	74	W	$[Xe]\,4f^{14}5d^46s^2$
23	V	$[Ar]\,3d^34s^2$	75	Re	$[Xe]\,4f^{14}5d^56s^2$
24	Cr	$[Ar]\,3d^54s^1$	76	Os	$[Xe]\,4f^{14}5d^66s^2$
25	Mn	$[Ar]\,3d^54s^2$	77	Ir	$[Xe]\,4f^{14}5d^76s^2$
26	Fe	$[Ar]\,3d^64s^2$	78	Pt	$[Xe]\,4f^{14}5d^96s^1$
27	Co	$[Ar]\,3d^74s^2$	79	Au	$[Xe]\,4f^{14}5d^{10}6s^1$
28	Ni	$[Ar]\,3d^84s^2$	80	Hg	$[Xe]\,4f^{14}5d^{10}6s^2$
29	Cu	$[Ar]\,3d^{10}4s^1$	81	Tl	$[Xe]\,4f^{14}5d^{10}6s^26p^1$
30	Zn	$[Ar]\,3d^{10}4s^2$	82	Pb	$[Xe]\,4f^{14}5d^{10}6s^26p^2$
31	Ga	$[Ar]\,3d^{10}4s^24p^1$	83	Bi	$[Xe]\,4f^{14}5d^{10}6s^26p^3$
32	Ge	$[Ar]\,3d^{10}4s^24p^2$	84	Po	$[Xe]\,4f^{14}5d^{10}6s^26p^4$
33	As	$[Ar]\,3d^{10}4s^24p^3$	85	At	$[Xe]\,4f^{14}5d^{10}6s^26p^5$
34	Se	$[Ar]\,3d^{10}4s^24p^4$	86	Rn	$[Xe]\,4f^{14}5d^{10}6s^26p^6$
35	Br	$[Ar]\,3d^{10}4s^24p^5$	87	Fr	$[Rn] \quad 7s^1$
36	Kr	$[Ar]\,3d^{10}4s^24p^6$	88	Ra	$[Rn] \quad 7s^2$
37	Rb	$[Kr]\,5s^1$	89	Ac	$[Rn]\,6d^17s^2$
38	Sr	$[Kr]\,5s^2$	90	Th	$[Rn]\,6d^27s^2$
39	Y	$[Kr]\,4d^15s^2$	91	Pa	$[Rn]\,5f^26d^17s^2$
40	Zr	$[Kr]\,4d^25s^2$	92	U	$[Rn]\,5f^36d^17s^2$
41	Nb	$[Kr]\,4d^45s^1$	93	Np	$[Rn]\,5f^46d^17s^2$
42	Mo	$[Kr]\,4d^55s^1$	94	Pu	$[Rn]\,5f^6 \quad 7s^2$
43	Tc	$[Kr]\,4d^65s^1$	95	Am	$[Rn]\,5f^7 \quad 7s^2$
44	Ru	$[Kr]\,4d^75s^1$	96	Cm	$[Rn]\,5f^76d^17s^2$
45	Rh	$[Kr]\,4d^85s^1$	97	Bk	$[Rn]\,5f^9 \quad 7s^2$
46	Pd	$[Kr]\,4d^{10}$	98	Cf	$[Rn]\,5f^{10} \quad 7s^2$
47	Ag	$[Kr]\,4d^{10}5s^1$	99	Es	$[Rn]\,5f^{11} \quad 7s^2$
48	Cd	$[Kr]\,4d^{10}5s^2$	100	Fm	$[Rn]\,5f^{12} \quad 7s^2$
49	In	$[Kr]\,4d^{10}5s^25p^1$	101	Md	$[Rn]\,5f^{13} \quad 7s^2$
50	Sn	$[Kr]\,4d^{10}5s^25p^2$	102	No	$[Rn]\,5f^{14} \quad 7s^2$
51	Sb	$[Kr]\,4d^{10}5s^25p^3$	103	Lr	$[Rn]\,5f^{14}6d^17s^2$
52	Te	$[Kr]\,4d^{10}5s^25p^4$			

One of the earliest models of the arrangement of the electrons in molecules is that published by G. N. Lewis in 1916, although it had been used by him in teaching for a number of years before. Chemists have found this model so convenient that it is still today the most widely used simple model. Lewis represented the core of an atom by its symbol and the valence-shell electrons by the appropriate number of dots. Lewis dot diagrams for the main-group elements of the first six periods are given in Figure 1.1.

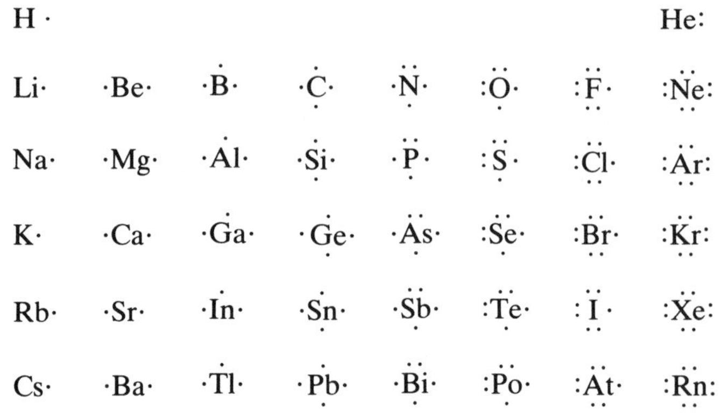

H ·   He:

Li·   ·Be·   ·B·   ·C·   ·N·   :O·   :F·   :Ne:

Na·   ·Mg·   ·Al·   ·Si·   ·P·   :S·   :Cl·   :Ar:

K·   ·Ca·   ·Ga·   ·Ge·   ·As·   :Se·   :Br·   :Kr:

Rb·   ·Sr·   ·In·   ·Sn·   ·Sb·   :Te·   :I·   :Xe:

Cs·   ·Ba·   ·Tl·   ·Pb·   ·Bi·   :Po·   :At·   :Rn:

Figure 1.1 Lewis symbols for the main-group elements.

Lewis proposed that in compound formation atoms achieve noble gas electron configurations either by electron loss or gain or by the sharing of one or more electron pairs. Because each of the noble gases, except helium, has eight electrons in its outer or valence shell, Lewis's proposal is often called the *octet rule*. Electrons are readily removed from the valence shell of an atom of a metal to give an ion with a noble gas electron configuration; for example,

$$Na(1s^2 2s^2 2p^6 3s^1) \longrightarrow Na^+(1s^2 2s^2 2p^6) + e^-$$

Nonmetals tend to gain electrons to give negative ions that have a noble gas configuration; for example,

$$O(1s^2 2s^2 2p^4) + 2e^- \longrightarrow O^{2-}(1s^2 2s^2 2p^6)$$

The positive and negative ions thus formed are attracted to each other by electrostatic forces to form ionic compounds, and such compounds are said to have *ionic bonds*.

Lewis also proposed that atoms may be held together by sharing one or more electron pairs. A shared electron pair constitutes a single *covalent bond* between the atoms, as in the Cl_2 molecule:

:Cl:Cl:

Similar Lewis diagrams can be written for many molecules in which each of the atoms has an octet of electrons in its valence shell. For example, the Lewis diagrams for carbon tetrafluoride, nitrogen trifluoride, oxygen difluoride, and fluorine are

$$\begin{array}{c} :\ddot{F}: \\ :\ddot{F}:\overset{\cdot\cdot}{C}:\ddot{F}: \\ :\ddot{F}: \end{array} \qquad \begin{array}{c} :\ddot{F}:\overset{\cdot\cdot}{N}:\ddot{F}: \\ :\ddot{F}: \end{array} \qquad \begin{array}{c} :\ddot{F}:\overset{\cdot\cdot}{O}: \\ :\ddot{F}: \end{array} \qquad :\ddot{F}:\ddot{F}:$$

In a completed valence shell all the electrons may be considered to be arranged in pairs, either *bonding pairs* or *nonbonding pairs*. Nonbonding pairs are also frequently called *lone pairs* or *unshared pairs*. Lewis diagrams are usually simplified by representing each bonding pair of electrons, that is, each covalent bond, by a single line, as follows:

$$\begin{array}{c} :\ddot{F}: \\ | \\ :\ddot{F}-\overset{\cdot\cdot}{C}-\ddot{F}: \\ | \\ :\ddot{F}: \end{array} \qquad \begin{array}{c} :\ddot{F}-\ddot{N}-\ddot{F}: \\ | \\ :\ddot{F}: \end{array} \qquad \begin{array}{c} :\ddot{O}-\ddot{F}: \\ | \\ :\ddot{F}: \end{array} \qquad :\ddot{F}-\ddot{F}:$$

Hydrogen is an exception to the octet rule in that the corresponding noble gas, helium, has only two electrons in its valence shell. Thus in the Lewis diagrams of compounds of hydrogen there are only two electrons in the valence shell of each hydrogen, as, for example, in methane, ammonia, water, and hydrogen fluoride:

$$\begin{array}{c} H \\ | \\ H-C-H \\ | \\ H \end{array} \qquad \begin{array}{c} \overset{\cdot\cdot}{N} \\ H-N-H \\ | \\ H \end{array} \qquad \begin{array}{c} H-\ddot{O}: \\ | \\ H \end{array} \qquad H-\ddot{F}:$$

In some molecules two or three electron pairs may be shared between two atoms, thereby forming double and triple bonds. Double bonds are found, for example, in ethene and carbon dioxide:

$$\begin{array}{c} H \\ \diagdown \\ \diagup \\ H \end{array} C{=}C \begin{array}{c} H \\ \diagup \\ \diagdown \\ H \end{array} \qquad :\ddot{O}{=}C{=}\ddot{O}:$$

Triple bonds are found for example in ethyne and hydrogen cyanide:

$$H-C{\equiv}C-H \qquad H-C{\equiv}N:$$

Although in the vast majority of covalent compounds atoms are held together by one or more shared pairs of electrons, there are many exceptions to the octet rule. For example, the central atoms in $BeCl_2(g)$ and in BCl_3 have only two and three electron pairs, respectively, in their valence shells:

$$Cl-Be-Cl \qquad \begin{array}{c} Cl-B-Cl \\ | \\ Cl \end{array}$$

There are also many molecules in which a central nonmetal atom from period 3 and beyond in the periodic table has five, six, or even more electron pairs in its valence shell. Examples include sulfur tetrafluoride, SF_4, chlorine trifluoride, ClF_3, and sulfur hexafluoride, SF_6:

$$
\begin{array}{ccc}
\text{F} & \text{F} & \text{F} \\
\vdots\,\text{S} & \vdots\ddot{\text{Cl}}\!-\!\text{F} & \text{S}
\end{array}
$$

IONIC AND COVALENT BONDS AND ELECTRONEGATIVITY

The concept of the covalent bond as consisting of a pair of shared electrons led to a great step forward in the understanding of chemical bonding and molecular structure, but Lewis could not explain *why* electrons are arranged in pairs nor why two shared electrons could bind two nuclei together. A full understanding of the covalent bond had to await the development of quantum mechanics, as we will discuss in Chapters 3 and 7. We will see that in the formation of a covalent bond there is a small increase in the electron density in the region between the atomic cores, and it is the electrostatic attraction between this increased electron density and the positively charged atomic cores that holds the two atomic cores together.

No sharp distinction can be drawn between covalent and ionic bonds, and in many molecules the bonds have an intermediate character. In fact, a pure covalent bond is only possible between atoms of the same kind, that is, atoms of the same element. Atoms of different elements attract the bonding electron pair to different extents, so there is some transfer of electron density from one atom to the other; that is, the bond is not purely covalent but has some ionic character. Similarly, there can be no pure ionic bond, because whenever two oppositely charged ions are attracted together there is inevitably at least a small amount of sharing of electron density between them. Molecules containing predominately ionic bonds are not common because normally, when oppositely charged ions are attracted together, they form a crystalline solid that consists of a regular periodic arrangement of ions in which no discrete molecules can be recognized (Figure 1.2). The whole crystal may be regarded as one giant molecule or as a polymer of essentially infinite size. Small molecules in which the bonding is predominately ionic are usually found only in the gas phase when an ionic solid is vaporized at high temperature.

The ability of an atom to attract the electrons of a covalent bond is called its *electronegativity*. Values of the electronegativities of the elements have been

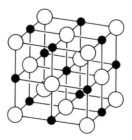

Figure 1.2 Structure of sodium chloride.

obtained by different authors by several different methods. Table 1.2 gives the values determined by A. L. Allred and E. G. Rochow. These can only be regarded as approximate average values because the electronegativity of an atom depends to some extent on the atoms that are attached to it and its oxidation state.

TABLE 1.2 ELECTRONEGATIVITY VALUES

Period	Group							
	1	2	3	4	5	6	7	8
1	H							He
	2.2							—
2	Li	Be	B	C	N	O	F	Ne
	1.0	1.5	2.0	2.5	3.1	3.5	4.1	—
3	Na	Mg	Al	Si	P	S	Cl	Ar
	1.0	1.2	1.3	1.7	2.1	2.4	2.8	—
4	K	Ca	Ga	Ge	As	Se	Br	Kr
	0.9	1.0	1.8	2.0	2.2	2.5	2.7	3.1
5	Rb	Sr	In	Sn	Sb	Te	I	Xe
	0.9	1.0	1.5	1.7	1.8	2.0	2.2	2.4

We also recognize a third type of bond, the metallic bond, in which a large number of atomic cores are held together by a delocalized electron cloud, and no discrete bonds between individual pairs of atoms and no molecules can be distinguished. Again, no sharp distinction can be drawn between metallic bonds and covalent and ionic bonds, and in many compounds the bonds have an intermediate character. However, since metallic bonding is a property of a large assembly of atoms, it is not found in the relatively small molecules with which we are primarily concerned in this book.

Both ionic bonding and metallic bonding are nondirectional in nature, and so the structures of ionic and metallic substances are determined primarily by the ways in which ions and atoms of different sizes may be packed together. In contrast, covalent bonds are highly directional in character, and it is the tendency for atoms to form bonds in specific directions that is the most important factor in determining the structures of covalent molecules and crystals.

THE VSEPR MODEL

The discussion of molecular geometry in this book is based on the valence-shell electron-pair repulsion (VSEPR) model, which in turn is based on Lewis's description of the electron arrangement in a molecule. The basic assumption of the VSEPR model is that the electron pairs in the valence shell of an atom, both bonding and nonbonding, adopt that arrangement that keeps them as far apart as possible; that is, they behave as if they repel each other. From the arrangement of the electron pairs in the valence shell of an atom the geometry of the covalent bonds that it forms can be easily predicted. In this book we are therefore mainly concerned with covalent molecules.

POLYHEDRA

Some simple molecules are linear and some others are planar, but most molecules are three dimensional. It is particuarly convenient to describe the structures of many three-dimensional molecules in terms of an appropriate polyhedron. So we now describe some of the more important polyhedra.

A *polyhedron* encloses a portion of three-dimensional space with four or more polygons. A *polygon* encloses a portion of a plane with three or more straight lines. A polygon is convex if each interior angle is less than 180°, and it is regular if it has equal interior angles and equal sides. In principle, there is an infinite number of regular polygons with the circle as a limiting case (Figure 1.3).

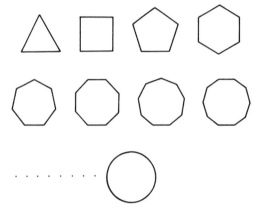

Figure 1.3 Regular polygons.

A polyhedron is convex if every *dihedral angle* is less than 180°. A dihedral angle is formed by two polygons joined along a common edge. A convex polyhedron is regular if its faces are equal regular polygons and if each of its vertices has the same surroundings.

There are only five regular convex polyhedra: the tetrahedron, cube (hexahedron), octahedron, dodecahedron, and icosahedron. They are shown in Figure 1.4. Their characteristic parameters are given in Table 1.3. The regular convex polyhedra are‚ called Platonic solids because they were an important part of Plato's natural philosophy.

The same polyhedron may be used to describe the shapes of two rather different types of molecule. For example, As_4 and B_4Cl_4 are described as tetrahedral molecules. In both these cases, there are atoms at the corners of a tetrahedron, and each of these atoms is bonded to each of its three neighbors so that there are bonds along all six edges of the tetrahedron (Figure 1.5). In contrast, methane, CH_4, and silane, SiH_4, are also both described as tetrahedral molecules. In these molecules there are atoms at each of the corners of the tetrahedron, but these atoms are not bonded together; rather they are bonded to an atom at the center of the tetrahedron (Figure 1.6). To distinguish these two types of tetrahedral molecules, we will call the latter type *centered* tetrahedral

tetrahedron

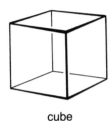

cube

dodecahedron

octahedron

icosahedron

Figure 1.4 The five regular polyhedra (Platonic solids).

TABLE 1.3 CHARACTERISTIC PARAMETERS OF THE REGULAR POLYHEDRA

Name	Total Number of Edges	Shape of Faces[a]	Number of Faces	Number of Edges Meeting at a Vertex	Number of Vertices
Tetrahedron	6	3	4	3	4
Cube	12	4	6	3	8
Octahedron	12	3	8	4	6
Dodecahedron	30	5	12	3	20
Icosahedron	30	3	20	5	12

[a]The numbers 3, 4, and 5 refer to equilateral triangle, square, and regular pentagon, respectively.

molecules. The other type of tetrahedral molecule in which the atoms are at the corners of a polyhedron and there is no central atom are often described as *cage* or *cluster* molecules. An example of an octahedral molecule without a central atom is the borane $B_6H_6^{2-}$ (Figure 1.7). There are many octahedral molecules with a central atom. Examples are SF_6 and PCl_6^- (Figure 1.7).

The regular convex polyhedra are highly symmetrical. There are various other families of less symmetrical polyhedra. The *semiregular polyhedra* are

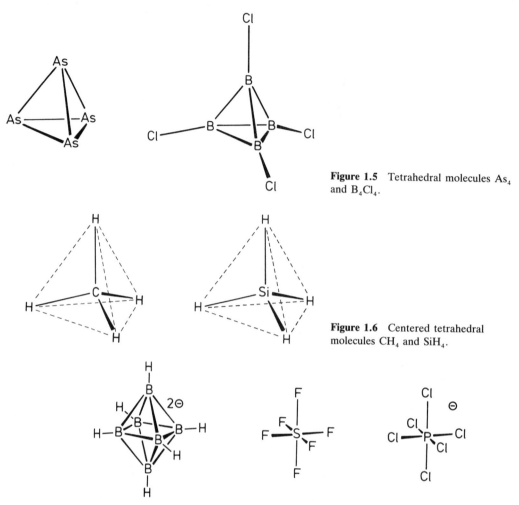

Figure 1.5 Tetrahedral molecules As_4 and B_4Cl_4.

Figure 1.6 Centered tetrahedral molecules CH_4 and SiH_4.

Figure 1.7 Octahedral molecules $B_6H_6^{2-}$, SF_6, PCl_6^-.

similar to the Platonic solids in that all their faces are regular and all their vertices have the same surroundings, but their faces are not all polygons of the same kind.

Prisms and *antiprisms* are also important polyhedra. A prism has two identical and parallel faces that are joined by a set of parallelograms. An antiprism also has two identical and parallel faces, but they are joined by a set of triangles. There is an infinite number of prisms and antiprisms and some of them are shown in Figure 1.8. The cube is a special case of a prism in which the two parallel faces are squares that are also joined by squares, and the octahedron is a special case of an antiprism in which the two parallel faces are equilateral triangles that are also joined by equilateral triangles.

Of the endless number of less regular polyhedra, two more kinds deserve special mention, the *pyramids* and *bipyramids*. The most symmetrical pyramids and bipyramids have a regular base. The tetrahedron is a special case of the

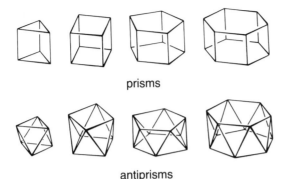

prisms

antiprisms

Figure 1.8 Prisms and antiprisms.

trigonal pyramid with a regular triangular base and three regular triangles as side faces. A bipyramid is a double pyramid, which may be thought of as obtained from a single pyramid by reflection with respect to its base. Examples are shown in Figure 1.9. A trigonal bipyramid is not a regular polyhedron as it has two sets

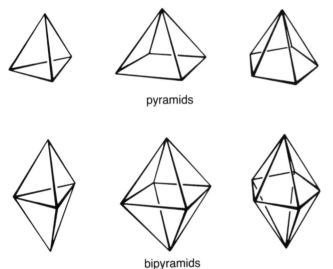

pyramids

bipyramids

Figure 1.9 Pyramids and bipyramids.

of nonequivalent vertices, at two of which three edges meet and at the other three four edges meet. The tetragonal bipyramid, however, becomes an octahedron when all its faces are regular triangles. Among the pyramids and bipyramids, the square pyramid and the trigonal bipyramid are important molecular shapes (Figure 1.10).

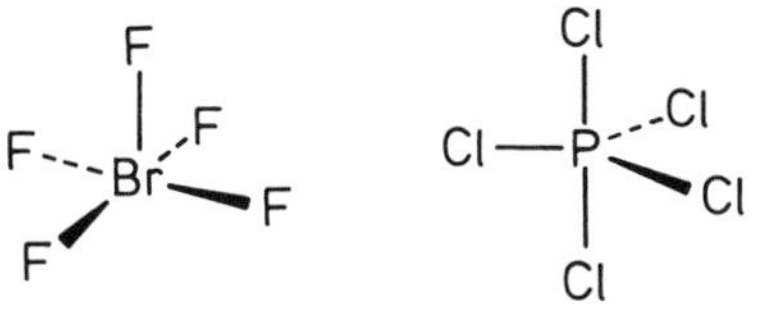

Figure 1.10 Square pyramidal (BrF_5) and trigonal bipyramidal (PCl_5) molecules.

The boranes and carboranes and the cage hydrocarbons $(CH)_n$ provide some nice examples of polyhedral molecules (Figure 1.11).

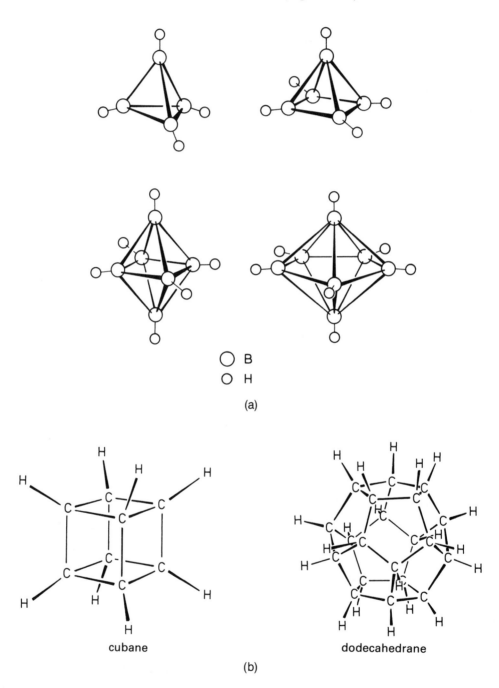

Figure 1.11 Cage molecules: (a) boranes and (b) hydrocarbons.

Ideal Bond Angles

When a molecule has the shape of a regular polygon or polyhedron, its bond angles are unambiguously defined. For example, a three-atom regular triangular molecule has 60° bond angles and a four-atom regular triangular molecule with an atom at the center of the triangle has 120° bond angles. A four-atom square molecule has 90° bond angles as does a five-atom molecule with an atom in the middle of the square (Figure 1.12).

Tetrahedral molecules have either 60° or 109.5° bond angles (Figure 1.13).

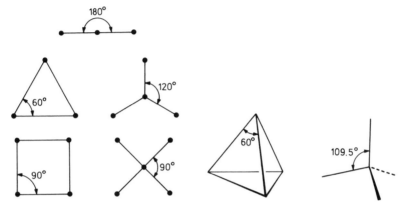

Figure 1.12 Ideal bond angles in linear, triangular, and square molecules.

Figure 1.13 Bond angles in tetrahedral molecules.

The value of 109.5° is given by the expression 2 arc sin $\sqrt{\frac{2}{3}}$, which can be derived using the Pythagorean theorem, as shown in Figure 1.14.

All the bond angles in a centered octahedral molecule are 90°, and in an octahedral molecule without a central atom they are 60° and 90° (Figure 1.15). A

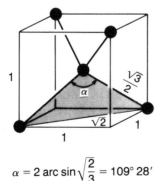

$$\alpha = 2 \text{ arc sin} \sqrt{\frac{2}{3}} = 109° \, 28'$$

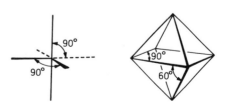

Figure 1.14 Derivation of the tetrahedral bond angle.

Figure 1.15 Bond angles in octahedral molecules.

trigonal bipyramidal molecule with a central atom has 90° angles between the axial and equatorial bonds and 120° angles between the equatorial bonds (Figure 1.16).

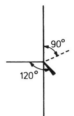

Figure 1.16   Bond angles in trigonal bipyramidal molecules.

BOND LENGTHS AND OTHER BOND PROPERTIES

The single most characteristic property of a chemical bond between two atoms is its length. The *bond length* is the distance between the nuclei of the two atoms that are bonded together. Molecular dimensions are most commonly quoted in ångströms (Å), picometers (pm), or nanometers (nm):

$$1\,\text{pm} = 1 \times 10^{-12}\,\text{m}$$

$$1\,\text{Å} = 1 \times 10^{-10}\,\text{m}$$

$$1\,\text{nm} = 1 \times 10^{-9}\,\text{m}$$

Because most bonds have lengths between 1 and 3 Å the ångström is a very convenient unit, and in the past it has been the most widely used unit for molecular dimensions. But the ångström is not part of the *Systeme Internationale*, and since the adoption of SI units the nanometer and the picometer are being increasingly used in the literature, especially in textbooks. Because of its convenience and its increasing popularity, the picometer is used for expressing bond lengths in this book. Bond lengths in ångströms can be converted to lengths in picometers by multiplying by 100; 1 Å = 100 pm.

Covalent Radii

The length of a single covalent bond between two given atoms often varies rather little from one molecule to another, and it is possible to divide the bond length into a contribution from each atom that is known as the *covalent radius* of the atom. This radius is sometimes also called the *atomic radius*, but the term covalent radius is to be preferred as this makes it clear that it refers to an atom forming a covalent bond and not to the free atom. Some single-bond covalent radii are listed in Table 1.4.

For many elements the covalent radius is obtained most simply by taking one-half of the length of the bond between the same two atoms. For example, the

TABLE 1.4 COVALENT RADII (pm)

H							He
37							—
Li	Be	B	C	N	O	F	Ne
152	111	88	77	70	66	64	—
Na	Mg	Al	Si	P	S	Cl	Ar
186	160	143	117	110	104	99	—
K	Ca	Ga	Ge	As	Se	Br	Kr
231	197	122	122	121	117	114	111
Rb	Sr	In	Sn	Sb	Te	I	Xe
244	215	162	140	141	135	133	130

covalent radius of chlorine is obtained from the bond length of the diatomic Cl_2 molecule:

$$r(Cl) = \tfrac{1}{2}d(Cl—Cl) = \tfrac{1}{2} \times 198 \text{ pm} = 99 \text{ pm}$$

And the covalent radius of carbon can be obtained from the C—C bond length in diamond:

$$r(C) = \tfrac{1}{2}d(C—C) = \tfrac{1}{2} \times 154 \text{ pm} = 77 \text{ pm}$$

To a good approximation, these covalent radii are additive. For example,

$$d(C—Cl) = r(C) + r(Cl) = 99 + 77 \text{ pm} = 176 \text{ pm}$$

which compares well with the experimentally determined value of 176.3(3) pm in CCl_4. However, for some atoms, notably F, O, and N, this method gives covalent radii that appear to be too large in that they predict values for most other M—F, M—O, and M—N bonds that are too long. The best that can be done is to select a value for a given element that gives the best agreement with the observed lengths of as large as possible a number of bonds formed by the element. It should be borne in mind that the covalent radius of an atom is an empirical parameter that varies somewhat with the oxidation state of the atom and the number and nature of the attached ligands. Nevertheless, it is often profitable to consider possible explanations when an observed bond length differs markedly from the value calculated from covalent radii.

Bond Order

The length of a bond between the same two atoms varies with the order of the bond. A single bond is said to have an order of 1, a double bond an order of 2, and a triple bond an order of 3. The covalent radii discussed above and given in Table 1.4 are for single bonds. In general, double bonds are shorter than single bonds and triple bonds are shorter still. The force of attraction exerted by the electrons of the bond for the two positive atomic cores is greater for the two electron pairs of a double bond than for a single electron pair, and greater still for three electron pairs. Covalent radii for atoms forming double and triple bonds can be obtained in the same way discussed above for single bonds. Some double and triple bond covalent radii are given in Table 1.5.

TABLE 1.5 SINGLE-, DOUBLE-, AND
TRIPLE-BOND RADII (pm)

Radius	C	N	O	P	S
Single bond	77	70	66	110	104
Double bond	67	60	56	100	94
Triple bond	60	55	52	—	—

Resonance Structures

The Lewis structure for the benzene molecule is

in which the CC bonds are alternately single and double and should therefore
have lengths of 154 pm and 134 pm, respectively. But the CC bonds are found
experimentally to all have the same length of 140 pm, which is intermediate
between the CC single- and double-bond lengths. The reason for these equal
bond lengths is that electrons are not always as localized as the Lewis structure
implies. Thus we can write two different but equivalent Lewis structures for
benzene:

According to these two structures a given CC bond may be considered to be
either a single bond or a double bond. This implies that there are three electron
pairs that cannot be considered to be localized between three particular pairs of
carbon atoms but must be considered to be delocalized over all six bonding
regions. These two structures are called *resonance structures*. The actual structure
in which the electrons are less localized than in either of these two structures may
be considered to be intermediate between these two structures, and thus each
bond has a bond order of 1.5, which is consistent with the observed bond length.

There are many other molecules in which at least some of the electrons are
less localized than is implied by a single Lewis structure and that are therefore

represented by two or more Lewis structures. For example, the carbonate ion in which all three bonds have the same length may be represented by the following three resonance structures:

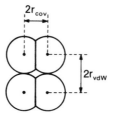

In this case each CO bond has a bond order of 1.33.

Nonbonded Distances

Molecular geometry is usually described in terms of bond lengths and bond angles. However, the distances between atoms that are not bonded together and the size of an atom in directions in which it is not bonded to other atoms are also frequently of interest. Such nonbonded distances are used to describe the size of a group of atoms or a molecule, which is important in discussing how different parts of a molecule interact or how molecules pack together. Nonbonded distances are most simply obtained from crystal structures of simple molecules.

Figure 1.17 shows a simplified schematic drawing of the packing of Cl_2 molecules in the solid state. The bond length is the distance between the two Cl nuclei in the same molecule. The nonbonded distance is the distance between two Cl nuclei in adjacent molecules. One-half the bond distance is taken as the covalent radius of chlorine. One-half of the nonbonded distance is called the *van der Waals* radius of chlorine. Whereas the covalent radius is the radius of an atom in the direction of a bond, the van der Waals radius is its radius in any other direction. Because an atom shares its electron density in the bond direction but not in other directions, the covalent radius is always smaller than the van der Waals radius of an atom. Like the covalent radius, the van der Waals radius of an atom cannot be assigned a precise value but only an approximate average value.

Figure 1.17 Van der Waals and covalent radii in $Cl_2(s)$.

Molecules are somewhat soft, and they can be pushed more or less closely together depending, for example on the structure of the crystal in which they are packed. Van der Waals radii of some elements are given in Table 1.6.

It is often observed that the distances between two atoms 1 and 3, separated by another atom 2, remains constant in a series of molecules, despite substantial changes in bond lengths and bond angles in the three-atom group (Figure 1.18). These *intramolecular 1, 3 interactions* can be considered to be a special case of van

TABLE 1.6 COVALENT (ATOMIC) RADII, INTRAMOLECULAR
1, 3 NONBONDED RADII, AND VAN DER WAALS RADII (pm)

Element	Covalent Radii	1, 3 Nonbonded Radii[a]	Van der Wals Radii[b]
B	88	133	—
C	77	125	—
N	70	114	150
O	66	113	140
F	64	108	135
Al	143	166	—
Si	117	155	—
P	110	145	190
S	104	145	185
Cl	99	144	180
Ga	122	172	—
Ge	122	158	—
As	121	161	200
Se	117	158	200
Br	114	159	195
In	162	195	—
Sn	140	188	—
Sb	141	188	220
Te	135	187	220
I	133	186	215

[a]L. S. Bartell, *J. Chem. Phys.*, **32**, 827 (1960). C. Glidewell,
Inorg. Chim. Acta, **20**, 113 (1976).
[b]L. Pauling, *The Nature of the Chemical Bond*, 3rd ed., Cornell
University Press, Ithaca, N.Y., 1960.

Figure 1.18 The intramolecular 1, 3
radius, $r_{1,3}$, of B in the angular
molecule AB_2.

der Waals interactions, and so they have also been called *intramolecular van der Waals interactions.* One-half of the 1, 3 nonbonded distance between two like atoms is the *1, 3 nonbonded radius* for the atom, and values for these radii have been proposed for a number of elements (Table 1.6). Like covalent and van der Waals radii, these 1, 3 nonbonded radii are only approximate average values. They are intermediate between the corresponding covalent and van der Waals radii.

Dissociation Energies

The shorter the bond is between atoms of the same two elements the stronger is the bond. However, there is no simple general relationship between the length and the strength of a bond. The most direct measure of the strength of a bond is the bond dissociation energy D. The general form of the relationship between the energy of interaction of two atoms that combine to form a bond and the distance between them is shown in Figure 1.19. The length of a bond corresponds to the position of the minimum of this potential energy function. The dissociation energy D is given by the depth of this potential energy function at the minimum. Some bond dissociation energies are listed in Table 1.7. These values were obtained from thermochemical and spectroscopic measurements.

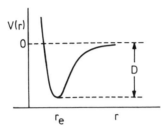

Figure 1.19 Potential energy (V) as a function of interatomic distance (r). D is the bond dissociation energy. The force constant is a measure of the steepness of the potential energy curve in the vicinity of the minimum.

TABLE 1.7 EXAMPLES OF BOND LENGTHS, BOND STRETCHING FORCE CONSTANTS, AND BOND DISSOCIATION ENERGIES

Bond	Bond Length (pm)	Force Constant (N/m)	Dissociation Energy (kJ/mol)
C—C	154	450	343
C=C	133	957	615
C≡C	120	1572	812
N—N	147	350	159
N=N	125	1300	418
N≡N	110	2240	946
C—N	147	490	293
C=N	125	1000	615
C≡N	115	1770	879

Force Constants

The force constant that is associated with the stretching vibration of a bond is often taken as a measure of the strength of the bond, although it is more correctly a measure of the rigidity of the bond. It is a measure of the steepness of the potential energy function around the minimum rather than the depth of the minimum (Figure 1.19). For a diatomic molecule the frequency of vibration v is determined by the force constant k and the reduced mass $\mu = m_1 m_2 / (m_1 + m_2)$, where m_1 and m_2 are the masses of the two atoms:

$$v = \frac{1}{2\pi} \sqrt{\frac{k}{\mu}}$$

The stronger the bond is, generally the higher the force constant and the higher the stretching frequency. For a polyatomic molecule the stretching force constant for each of the various bonds cannot in general be obtained in a completely unambiguous manner. Nevertheless, the values obtained, although often approximate, are useful as an approximate measure of bond strength or rigidity. Although there is no general relationship between the strength of a bond as measured by its dissociation energy and the rigidity of the bond as measured by its stretching force constant, in general the two parameters approximately parallel each other. Multiple bonds have higher dissociation energies and higher force constants than single bonds. Figure 1.20 shows the bond dissociation energy–bond order relationship for carbon–carbon bonds, and Table 1.7 lists some characteristic properties of single, double, and triple bonds.

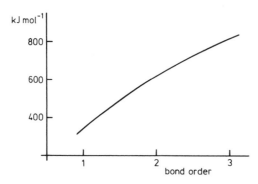

Figure 1.20 Dissociation energy–bond order relationship for carbon–carbon bonds.

ISOMERS

The chemical composition of a substance is expressed by its molecular formula. The particular pattern in which the atoms in a molecule are connected together by bonds is called the *atomic connectivity*. Two molecules are *homomers* if their formulas, their atomic connectivity, and the spatial arrangement of their nuclei are all the same. In other words, homomers are molecules that are superimposable (Figure 1.21). An important, although not unique, characteristic of the molecular geometry of homomers is the equality of all distances between corre-

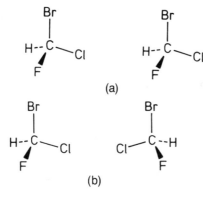

Figure 1.21 (a) Homomers are identical molecules. (b) Enantiomers differ from homomers only in that they are each other's mirror image. Enantiomers are said to be chiral.

sponding atoms. The equality of all distances is maintained in *enantiomers*, which differ from homomers only in that they are each other's mirror image (Figure 1.21), and they are therefore not superimposable. A molecule that has an enantiomer is said to be *chiral*. All other molecules are *achiral*.

Molecules with the same formula but in which the distances between corresponding atoms are not all the same are called *structural isomers*. Structural isomers are of two types. If their atomic connectivities are different, they are called *constitutional isomers* (Figure 1.22); if their atomic connectivities are the same, they are called *diastereomers*. Diastereomers can be further subdivided into two groups according to whether they can become superimposable by rotation about a bond or not. Isomers that can become superimposable by rotation about a bond are called *rotational isomers*. *Geometrical isomers* (Figure 1.23) and *con-*

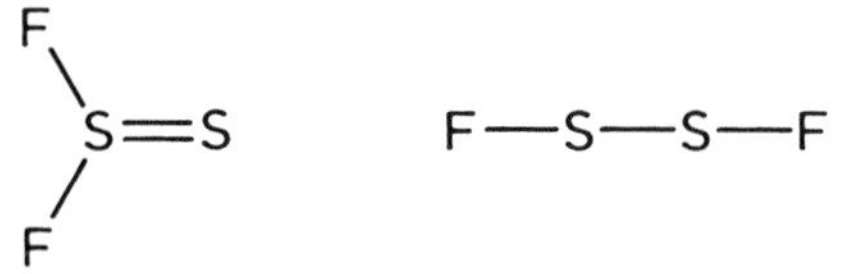

Figure 1.22 Constitutional isomers of S_2F_2. They have the same molecular formula but different atomic connectivity.

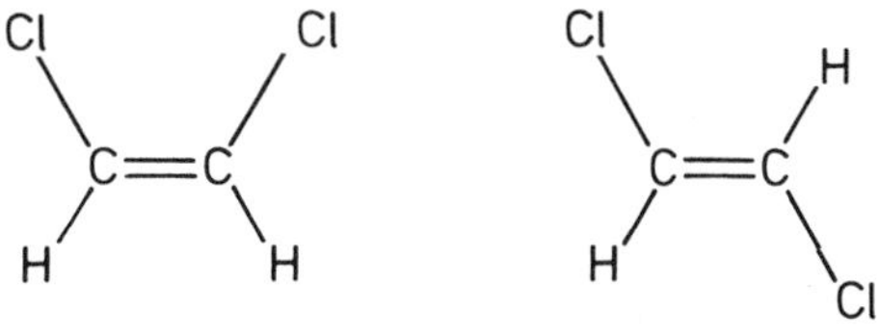

Figure 1.23 Geometrical isomers are rotational isomers for which the barrier to rotation is sufficiently high that they exist as separate isolatable species. The examples shown here are *cis* and *trans* 1,2-dichloroethene.

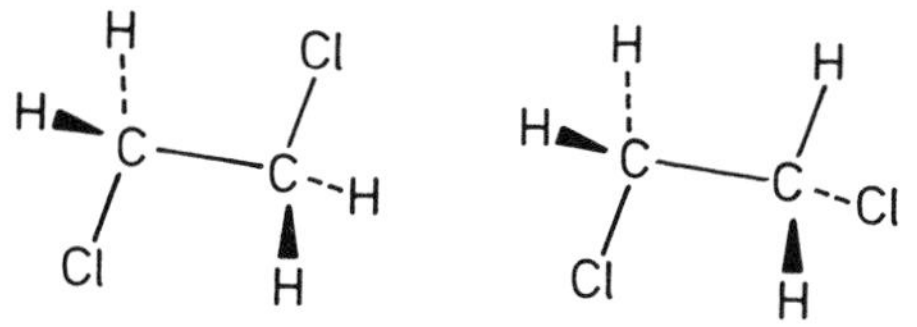

formers (Figure 1.24) are both rotational isomers. The difference between them is in the magnitude of the potential energy barrier separating the two forms. The barrier is so high for geometrical isomers (for example, a typical barrier is 300 kJ/mol) that the two forms can be physically separated. Thus geometrical

Figure 1.24 Conformers are rotational isomers for which the barrier to rotation is so low that they do not exist as separate isolatable species. Two conformers of 1,2-dichloroethane are shown here.

isomers can be considered to be truly different substances. On the other hand, a typical barrier for conformers is only a few kilojoules per mole so that their physical separation is not possible. If the torsional axis is a double bond about which the energy of rotation is high, the two isomers are geometrical isomers; but if the torsional axis is a single bond about which the energy of rotation is very low, the two isomers are conformers. Intermediate cases are also common. The relationships between the various types of isomers that we have discussed are summarized in Figure 1.25. Some of the names of the various types of isomers have a historical origin and are not very practical. For example, the name "geometrical isomerism" is clearly inadequate to distinguish it from other types of isomerism.

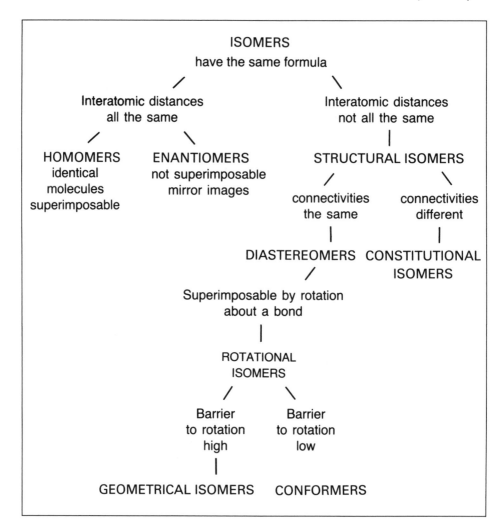

Figure 1.25 Relationships among the different kinds of isomers.

DYNAMIC STRUCTURES

In most of our discussion of molecular geometry throughout this book, we will consider the molecules to be rigid and motionless. However, we have to keep in mind that the molecules are permanently in motion. In addition to their translational motion, they are rotating and vibrating. The translational motion has no effect on molecular structure, and the slight centrifugal distortion due to rotation is not usually of any importance. However, the consequences of intramolecular vibrations cannot be entirely ignored when discussing the experimental determination of molecular geometry (see Chapter 2). In such a vibration, each nucleus in a molecule is displaced from its equilibrium position in a periodic manner.

Typically, there are about 10^{12} to 10^{14} vibrations per second. This vibrational motion occurs in all phases, including the crystal phase and at all temperatures down to the lowest possible temperature although its amplitude is greatly reduced at very low temperatures. The amplitude of molecular vibrations at room temperature amounts to several percent of the internuclear distance.

An extreme but not uncommon case of intramolecular motion is the exchange or permutation of the atomic nuclei within a molecule. This phenomenon was first observed in some trigonal bipyramidal molecules. In this case the atoms in the equatorial and axial positions rapidly exchange positions through an intermediate square pyramidal structure, as is illustrated in Figure 1.26. This

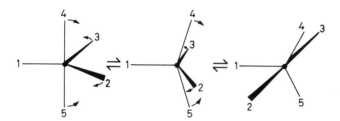

Figure 1.26 Permutational isomerization (pseudorotation) of a trigonal bipyramidal AX_5 molecule. The isomerization occurs through a square pyramidal transition state.

motion has been called pseudorotation because, after it has occurred, the newly formed trigonal bipyramidal molecule has the same orientation as if it had been rotated by 90°. Large-amplitude bending vibrations of the two axial atoms and two of the equatorial atoms bring them into the same plane, the four atoms then forming the base of a square pyramid with the third equatorial atom now in the apical position of the square pyramid. Further motion of these atoms results in the two axial atoms ending up in equatorial positions and the two originally equatorial atoms in the axial positions. Thus two of the equatorial positions have been permuted with the two axial positions. Continuation of this process rapidly exchanges the atoms through all the equatorial and axial positions.

We will see later that different experimental techniques for studying molecular structure utilize different physical phenomena in which the interaction times of radiation and matter may be very different. A relatively slow measurement will detect only a mean of the interconverting configurations, whereas measurements with interaction times shorter than the time period of the interconversion are capable of detecting distinct configurations. Thus it is important to examine the consequences of molecular vibrations with reference to the time scale of the measurement.

Although, in principle, permutation of the nuclei may occur in all molecules, while it happens billions of times a second in a PF_5 molecule, it may happen only once or twice in 10^{15} years in a methane molecule.

Another consequence of molecular vibrations is that a molecule may appear to have a different shape from its true equilibrium geometry. Figure 1.27 shows the example of a metal dihalide molecule that has a linear equilibrium geometry. However, bending vibrations may make it appear to have a bent V-shaped geometry. We will return to the consequences of intramolecular motions in the discussion of the experimental determination of molecular geometry in Chapter 2.

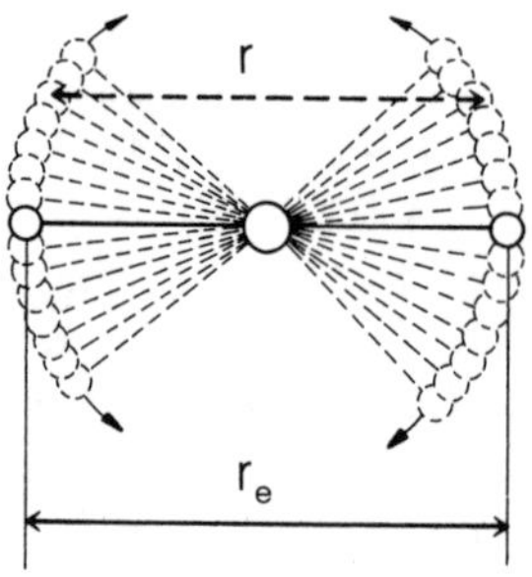

Figure 1.27 A linear triatomic molecule, MX$_2$, may appear to have a V-shape due to its bending vibration because the average distance X . . . X(r) is shorter than the equilibrium distance X . . . X(r_e).

REFERENCES AND SUGGESTED READING

R. J. Gillespie, D. A. Humphreys, N. C. Baird, and E. A. Robinson, *Chemistry*, 2nd Ed., Allyn and Bacon, Boston, 1989.

I. Hargittai and M. Hargittai, *Symmetry through the Eyes of a Chemist*, VCH, Weinheim, 1986; VCH, New York, 1987.

G. N. Lewis, *J. Amer. Chem. Soc.*, **38**, 762, 1916.

L. Pauling, *The Nature of the Chemical Bond*, 3rd Ed., Cornell University Press, Ithaca, N.Y., 1960.

_2_
The Determination of Molecular Geometry

The purpose of this chapter is to briefly describe the most important methods by which the geometry of a molecule may be determined, with particular emphasis on the type and accuracy of the information that can be obtained and the main limitations of each method.

The information that we can obtain about the geometry of a molecule is of two kinds, qualitative and quantitative. Some methods give only qualitative information on the general shape and symmetry of a molecule, while others give complete quantitative information on the relative positions of all the atoms in a molecule, which can be described conveniently by bond lengths, bond angles, and torsional angles.

The experimental methods can be grouped according to the nature of the interaction between matter and radiation in the physical phenomena upon which they are based. There are two main groups.

1. *Diffraction methods* are based on scattering phenomena in which a portion of the radiation changes direction without energy transfer. Diffraction effects arise from interference between the scattered radiation. This radiation may be electromagnetic, in particular X-rays, or very fast moving small particles, such as electrons or neutrons. X-ray diffraction and neutron diffraction are used for determining molecular structures in crystals, and electron diffraction is used for the determination of the structures of molecules in the gas phase. All three methods give quantitative information on molecular geometry.

2. *Spectroscopic methods* are based on those interactions that are accompanied by a transfer of energy between electromagnetic radiation and matter. Microwave spectroscopy gives quantitative information on small molecules in the gas phase, while vibrational spectroscopy (infrared and Raman spectroscopy) gives only qualitative information on molecular geometry in the gas, liquid, or solid phase.

X-RAY CRYSTALLOGRAPHY

When light falls on a regularly spaced set of slits, it is scattered in all directions by each slit and the resulting interference pattern depends on the wavelength of the light and the distance between the slits. This phenomenon is called *diffraction*. In an optical experiment the spacing of the slits is usually known, and the purpose of the experiment is to determine the wavelength of the light. In diffraction studies of crystal structures, the wavelength must be much smaller than the wavelength of visible light in order to be comparable with the distances between the atoms. X-rays have a suitable wavelength, and in an X-ray diffraction experiment the wavelength is known so that the interatomic distances can be determined.

X-rays are scattered by the electron cloud of an atom. The intensity of the X-rays scattered by an atom depends on the number of electrons in the atom, that is, on the atomic number Z. Because the atoms in a crystalline solid are arranged in a regular repeating pattern, interference between the X-rays scattered by the electron cloud of each atom gives rise to a diffraction pattern. In a modern diffractometer a beam of X-rays is directed onto a small crystal, and the crystal is rotated successively into all positions necessary to give a complete set of diffracted beams. From the intensities and directions of the scattered beams of X-rays, the three-dimensional pattern of electron density distribution in the crystal can be deduced (Figure 2.1).

Since the electron density is strongly concentrated around each nucleus, the position of the nucleus can usually be obtained with considerable accuracy, particularly for heavy atoms. For light atoms with few electrons, and most notably for hydrogen, there is only a small electron density even in the immediate vicinity of the nucleus, and the position of the nucleus cannot be determined with as much accuracy as for heavy atoms. Even if the electron density corresponding to a hydrogen atom can be located from an X-ray diffraction pattern, because it forms part of a covalent bond, this electron density is displaced from around the hydrogen nucleus toward the other nucleus with which it is bonded, so that the position of the hydrogen nucleus cannot be obtained with great accuracy.

Another limitation on the accuracy with which the position of an atom may be determined is a consequence of the vibrational motion of the atoms relative to each other and the librational motion of the molecule as a whole. Figure 2.2 illustrates the riding motion of a linear triatomic group. As a result of this motion, the electron density distribution of each of the two end atoms appears to be somewhat smeared out, and the center of gravity of this more diffuse distribution is closer to the pivotal atom than the center of gravity of the true electron

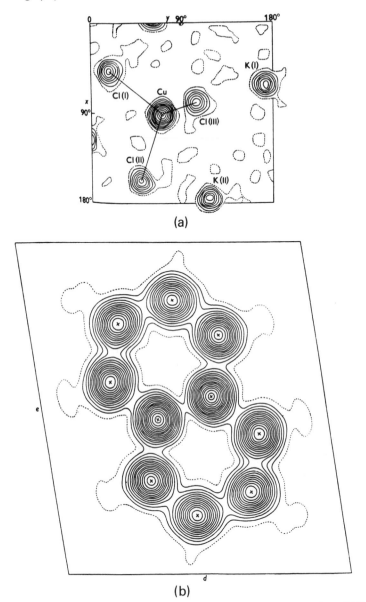

(a)

(b)

Figure 2.1 Electron density maps: (a) K_2CuCl_3 along a crystal face [001]. (Reproduced with permission from C. Brink and C. H. MacGillavry, *Acta Crystallogr.* 2, 1949, p. 158.) (b) Naphthalene, $C_{10}H_8$, through the plane of the molecule. (Reproduced with permission from S. C. Abrahams, J. M. Robertson and J. G. White, *Acta Crystallogr.* 2, 1949, *p*. 238.)

distribution. Thus the bonds appear to be shorter than the true internuclear distance. This effect can be minimized by working at low temperatures, and a suitable correction for the effect can usually be made. However, it is important to be aware that vibrational and librational motions can lead to errors in bond

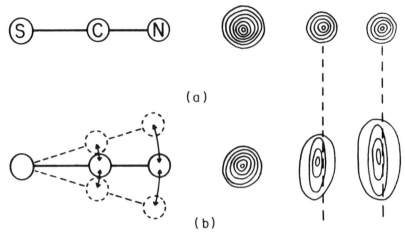

(a)

(b)

Figure 2.2 Diagram showing the effect of "riding" motion on a linear group: (a) No thermal vibration. (b) Thermal vibration normal to the axis of the group, showing that the mean electron density is displaced toward the pivot point. (Reproduced with permission from I. Hargittai and I. C. Paul, in S. Patai, ed., *The Chemistry of Cyanates and Their Thio Derivatives*, Wiley, Chichester, England, 1977, copyright (1977) John Wiley & Sons, Ltd.)

lengths, particularly if the necessary corrections are not made. The graphical representation of a structure determination is usually given not as an electron density distribution map but as a structural diagram in which the atoms are shown as ellipsoids, where the size and shape of the ellipsoid represent the relative amplitudes of the motion of the atom in each direction in space (Figure 2.3).

X-ray diffraction is primarily applied to crystalline solids, and it is not convenient for substances that are liquid or gaseous down to low temperatures. It must also be borne in mind that the geometry of a molecule in the solid state may be affected by interactions with its neighbors, so its geometry may differ somewhat from that of the free isolated molecule in the gas phase. In extreme cases the structure of a substance may be quite different in the solid state and in the gas phase. For example, in the gas phase phosphorus pentachloride consists of trigonal bipyramidal covalent PCl_5 molecules, but in the solid state it consists of tetrahedral PCl_4^+ ions and octahedral PCl_6^- ions.

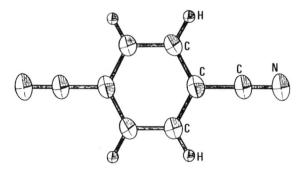

Figure 2.3 Graphical representation of the X-ray crystallographic structure of para-dicyanobenzene with atoms depicted as thermal ellipsoids. (Reproduced with permission from M. Colapietro, A. Domenicano, G. Portalone, G. Schultz, and I. Hargittai, *J. Mol. Struct.* 112, 1984, p. 141.)

NEUTRON DIFFRACTION

According to the quantum theory, moving particles can be characterized by a wavelength that depends on the mass and velocity of these particles. As a result of their wave properties, particles as well as electromagnetic radiation can be diffracted. However, only the lightest particles at high velocity have wavelengths that are comparable with the distances between atoms in molecules and crystalline solids, and in practice only electrons and neutrons are used to obtain diffraction patterns.

Neutrons are scattered directly by the atomic nuclei and are largely unaffected by the surrounding electrons. Morever, neutrons are scattered about as strongly by light atoms as by heavy atoms. Thus in neutron diffraction the positions of the nuclei of light atoms can be determined as accurately as for heavy atoms. This is a particular advantage for determining the positions of hydrogen atoms.

The biggest disadvantage of neutron diffraction is that a suitable beam of neutrons is not readily available. Because a neutron beam is usually produced by slowing down the fast neutrons obtained from a nuclear reactor, neutron diffraction experiments can only be carried out in a few places. Thus neutron diffraction is much less commonly used for structure determination than X-ray crystallography. Its most important application is for the determination of the positions of light atoms, such as hydrogen, for which it is the superior method.

ELECTRON DIFFRACTION

The interaction between matter and electrons is very strong, and for this reason electrons do not penetrate very far into a solid or a liquid. They are very suitable, however, for the determination of the structure of molecules in the gas phase. A beam of fast-moving electrons is obtained by accelerating them with an electric field. This beam of electrons is fired through a narrow stream of gas molecules. Although a gas has a low density, the electron beam may be made very intense, and as the interaction of electrons with a molecule is orders of magnitude stronger than the interaction of X-rays with the same molecule, it is possible to obtain a sufficiently intense diffraction pattern.

The negatively charged electrons are scattered because of their electrostatic interaction with the positively charged nucleus and negatively charged electrons of the molecule. The interaction with the nucleus is much stronger than with the electrons, since the nuclei represent a relatively large positive charge concentrated in a very small volume. Thus the relative positions of the nuclei can be determined directly as in neutron diffraction.

A diffraction pattern results from the interference of the radiation scattered by each of the atoms in the molecule. Because a gas consists of molecules of random orientations, its diffraction pattern consists of diffuse concentric rings. The diffraction pattern is superimposed on a background due simply to the scattering from the atoms. After subtracting the background and subjecting the intensity curve to the mathematical operation called Fourier transformation, a radial distribution curve is obtained. This curve has a peak for each of the interatomic distances in the molecule, and the area of each peak is proportional to the atomic numbers of the atoms involved and to the number of times that the particular interatomic distance occurs in the molecule (Figure 2.4). On the basis of this information it is usually possible to choose between alternative structures for the molecule, as shown, for example, for $OSF_2(CF_3)_2$ in Figure 2.5, by comparing radial distributions, calculated for different models, with the experimental radial distribution curve. If the agreement with one of the possible structures is reasonably good, the geometrical and vibrational parameters may be varied until the best possible agreement with the experimental curve is obtained.

An obvious limitation of the electron diffraction method is that it is restricted to substances that have a vapor pressure of at least 1 torr at a temperature at which they are stable. However, even such involatile species as alkali metal salts and metal oxides have been studied because they are stable at temperatures of 2000 K or more. The accuracy with which structural parameters involving light atoms can be determined by electron diffraction is intermediate between that of X-ray diffraction and of neutron diffraction. An important limitation is that the peaks for various interatomic distances appear separately on the radial distribution for the simplest molecules only. For larger molecules they overlap, and this may hinder their assignment.

Because the interpretation of the experimental data usually involves the comparison of radial distribution curves calculated for different assumed structures of the molecule with the experimental radial distribution, it is possible that a structure that was not considered could agree with the experimental data better than any of the structures that were considered. Incorrect structures have in fact

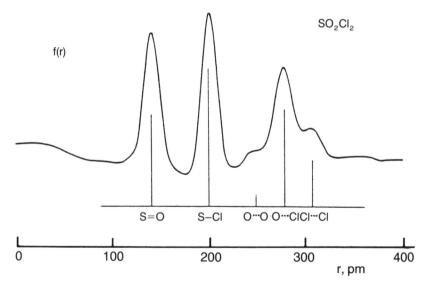

Figure 2.4   Electron diffraction radial distribution of sulfuryl chloride, SO_2Cl_2. The contributions of various atom pairs are indicated. (After M. Hargittai and I. Hargittai, *J. Mol. Struct.* 73, 1981, p. 253.)

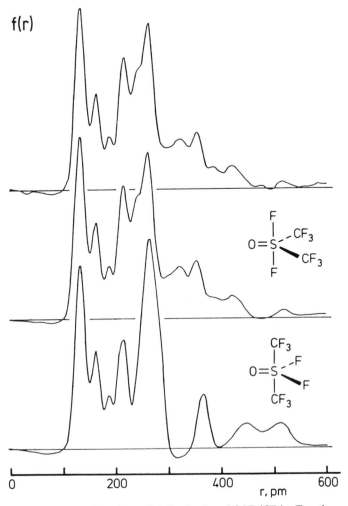

Figure 2.5 Electron diffraction radial distribution of $OSF_2(CF_3)_2$. Two theoretical curves, calculated for two models, are shown in addition to the experimental
distribution (top curve). Only one of the two models is in agreement with the
experimental data. (After H. Oberhammer, J. M. Shreeve, and G. L. Gard,
Inorg. Chem. 23, 1984, p. 2820, with permission, copyright (1984) American
Chemical Society.)

been deduced in this way in the past. In modern studies the reliability of
structural information from electron diffraction is greatly enhanced by using
complementary data from other methods.

ROTATIONAL SPECTROSCOPY

A rotating permanent dipole, such as a polar molecule in the gas phase, interacts
with electromagnetic radiation if the frequency of the radiation is commensurate
with the rotational frequencies of the dipole. Thus any rotating molecule that has

a permanent dipole moment absorbs energy at the appropriate frequencies and is excited to a higher rotational level in which it is rotating more rapidly. The rotational frequencies of a molecule depend on the three moments of inertia, I_a, I_b, and I_c, each of which is a sum of the products of the atomic masses m_i and the square of the distances r_i of the atoms from an appropriate axis:

$$I_i = \sum m_i r_i^2$$

The summation extends over all the atoms in the molecule. Depending on the symmetry of a molecule, it may have one, two, or three different values for the moments of inertia about three mutually perpendicular axes. There is a set of energy levels corresponding to each moment of inertia and as the molecule absorbs radation of the appropriate frequency it is promoted from one energy level to the next.

For very symmetric molecules such as CH_4, that are called *spherical tops*, all three moments of inertia are equal to each other. Another class of molecules, such as CH_3F, BF_3, and benzene, which have somewhat less symmetry and are called *symmetric tops*, has two of the three moments of inertia equal to each other. A linear molecule is a special case of this class because the unique moment of inertia is essentially zero. For most molecules (*asymmetric tops*), the three moments of inertia are different.

From the rotational spectrum of a molecule, its moments of inertia can be accurately determined. The rotational frequencies of molecules are of the order of 10^{13} to 10^9 Hz, so the rotational spectra of polar molecules are observed in the far infrared, millimeter wave, and microwave regions of the electromagnetic spectrum. For all except the lightest molecules the spectrum is in the microwave region, so this type of spectroscopy is usually called *microwave spectroscopy*.

For a diatomic molecule there is only one structural parameter, the bond length, and the single moment of inertia that can be obtained from the rotational spectrum is sufficient to determine this with considerable precision, assuming that the atomic masses are known with sufficient accuracy. For less symmetrical molecules, up to three moments of inertia can be obtained from the spectra, and hence up to three geometrical parameters may be determined. It would appear therefore that this method is limited to the determination of the structures of the simplest molecules only. However, the method can be extended to somewhat larger molecules by obtaining the spectra of isotopically substituted species and assuming that the bond lengths and bond angles are not significantly different in the isotopically substituted molecules. In this way, up to three more moments of inertia and therefore up to three more geometrical parameters may be obtained for each isotopic substitution.

The rotational spectrum of nonpolar molecules can also, in principle, be obtained from the rotational Raman spectrum, but mainly because of experimental difficulties this technique has not been widely used.

VIBRATIONAL SPECTROSCOPY

All molecules are undergoing vibrational motions. The possible modes of vibrational motion can be approximately described as bond stretching, bending (that is, bond angle deformation), and torsion. Displacement of the atoms during vibration leads to oscillations of the electric charge distribution in a molecule. Such an oscillating charge distribution interacts with the oscillating electric vector of electromagnetic radiation. If the frequency of the vibration is commensurate with the frequency of the electromagnetic radiation, a quantum of energy can be absorbed by the molecule, which raises it to a higher vibrational level in which the vibrational motion has a greater amplitude. The vibrational frequencies of a molecule are generally in the range of 10^{14} to 10^{12} Hz, and the corresponding electromagnetic frequencies are in the infrared region of the spectrum.*

The vibrational frequencies of a molecule may also be obtained from the Raman spectrum. When a molecule interacts with radiation of much higher frequency than its vibrational frequencies, it may absorb a quantum of radiation, thereby exciting the molecule to an excited electronic state, called a virtual state. This excited state has a very short lifetime, so it almost immediately reemits a photon whose energy may differ from the energy of the incident photon by a quantum of vibrational energy. The resulting spectrum is called the *Raman spectrum*: it contains lines whose frequencies differ from those of the incident radiation by the frequencies of the molecular vibrations. The interaction comes about through the polarization of the electron charge cloud of the molecule by the oscillating electric vector of the incident radiation. The important molecular property connected with the Raman spectrum is therefore the polarizability, and a vibration will give rise to a Raman line if it leads to a change in the polarizability of the molecule.

The most common use of the vibrational spectrum of a molecule is as a fingerprint, that is, for the identification of a substance by comparison of its vibrational spectrum with that of a previously recorded spectrum of the pure compound. Another important use is for the calculation of the force constants, which are related to bond strength (Chapter 1). But the vibrational frequencies do not give any quantitative information about the geometry of a molecule. However, the number of bands in the infrared and Raman spectra may give information about the symmetry and shape of the molecule.

The total number of modes of vibration of a nonlinear molecule is $3n - 6$, where n is the number of atoms in the molecule. Thus for the SF_4 molecule, which has five atoms, we expect a total of nine modes of vibrations. But, depending on the symmetry of the molecule, not all these will be observed in the

---

*Molecular vibrational frequencies are usually quoted in units of reciprocal centimeters or wave numbers rather than frequencies, that is, by the number of waves per centimeter:

$$\text{wave number (cm}^{-1}) = \frac{\text{frequency (s}^{-1})}{\text{speed of light (cm s}^{-1})}$$

Typical wave numbers are in the range from 50 to 3500 cm^{-1}.

infrared and Raman spectra because they do not all lead to a change either in the dipole moment or in the polarizability of the molecule. Thus, if SF_4 had a tetrahedral shape like CF_4, only two bands would be observed in the infrared spectrum and four in the Raman spectrum. If SF_4 had a trigonal pyramid shape, six fundamental vibrational frequencies would be observed in both the infrared and Raman spectra. If SF_4 had a disphenoid shape (see Figure 5.68), eight vibrational frequencies would be observed in the infrared spectrum and all nine in the Raman spectrum. Thus, in principle at least, it is possible to distinguish between these molecular shapes simply on the basis of the number of bands observed in the infrared and Raman spectra. In fact, five bands have been observed in the infrared spectrum and five in the Raman spectrum of SF_4. This observation clearly eliminates the tetrahedral structure, but because some bands were apparently too weak to be observed it does not distinguish between the disphenoid and trigonal pyramid shapes. Other information was used to show that the molecule has a disphenoid shape.

In general, the number of bands observed in the infrared and Raman spectra of a molecule increases with decreasing molecular symmetry. Thus, if the number of observed bands is greater than that predicted for a given symmetry, it can be concluded that the molecule must have a lower symmetry. But if for some reason the complete spectrum of a molecule is not obtained, incorrect conclusions concerning its shape may be reached. For example, an early study of $OClF_3$ found six vibrational frequencies and it was concluded that the molecule had a tetrahedral shape. However, a later study found three more fundamental frequencies, which showed that it must have a less symmetrical structure. It is now believed to have a disphenoid shape consistent with the number of observed bands in the infrared and Raman spectra.

A limitation of the method is that it is difficult to obtain good spectra for substances in the gas phase, particularly for those that have a low vapor pressure. Thus most measurements of vibrational spectra are made on liquids, solutions, and solids, and they therefore refer to molecules that may be interacting with each other or with solvent molecules, and not to free molecules. This difficulty can be largely overcome by using the matrix isolation technique in which the vapor of the substance to be studied is mixed with an inert gas, such as argon or krypton, and is then frozen to give a dilute solid solution of the substance in the inert gas. However, in a few cases interactions even with noble gases have been observed that cause shifts in the vibrational frequencies of the molecules.

DIPOLE MOMENTS

Whether or not a polyatomic molecule has a dipole moment depends on its shape. Thus a symmetrical triatomic molecule with polar bonds has a dipole moment if it is bent, but no dipole moment if it is linear. Thus dipole moments have been used for many years to obtain qualitative information about molecular shape. However, the determination of structure by means of dipole moment measurements has largely been superseded by other methods that give more information. But one particular method of determining dipole moments, the electric deflection of molecular beams, is of special importance. Only polar molecules are deflected by an inhomogeneous electric field; so if a beam of molecules is passed through an inhomogeneous electric field, the dipole moment of the molecules can be measured. This method has been used for obtaining information on the shapes of some simple molecules, such as metal halides, in the gas phase, which cannot be easily obtained by other methods.

NUCLEAR MAGNETIC RESONANCE

Nuclear magnetic resonance (NMR) spectroscopy has become a structural tool of great importance in chemistry during the past 20 years. However, the NMR spectra of substances in solution or in the gas phase only give information on the connectivity (or topology) of a molecule and not on its geometry, except in some cases in a very indirect manner. The differences in the resonance frequencies (called *chemical shift*) of atoms of the same kind enable us to determine which of these atoms are geometrically nonequivalent, and the fine structure in the resonance lines resulting from magnetic interaction between the atoms via the bonding electrons enables us to deduce which atoms are directly bonded together. Although magnetic nuclei interact through the bonding electrons to produce spin–spin coupling patterns, their direct through space (dipole–dipole) interaction is averaged to zero by the random motions of the molecules, so no information on the geometric parameters of the molecule can be obtained. However, in a liquid crystal solution all the molecules tend to align in one particular direction so that the direct through space couplings between nuclei can be observed and hence the distance between the nuclei can be found. However, this technique has until now only been used rather little, and there are some severe limitations to its applicability.

TIME SCALES

In the section on dynamic structures in Chapter 1, we alluded to the importance of the relative magnitudes of the interaction time of the physical measurement and the lifetime of a particular structure. By interaction time we mean the time during which radiation interacts with a molecule. For example, the interaction time in electron diffraction is the time during which the electrons are traveling in the vicinity of the molecule and during which they are therefore under the influence of its charge distribution. Thus this interaction time depends on the speed of the electron and on the size of the molecule. It is of the order of 10^{-18} s.

An example of the importance of the relative magnitudes of interaction times and lifetimes is provided by the trigonal bipyramidal molecule $(CH_3)_2NPF_4$ with the dimethylamino group in one of the equatorial positions. The ^{31}P NMR spectrum of this molecule changes markedly with temperature as shown in Figure 2.6. At low temperatures the spectrum shows the presence of nonequivalent fluorine atoms, that is, axial and equatorial fluorines, respectively. At this

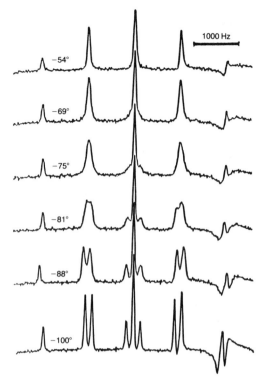

Figure 2.6 ^{31}P NMR spectra of $(CH_3)_2NPF_4$ as a function of the temperature. (Reproduced with permission from G. M. Whiteside and H. Lee Mitchell, *J. Am. Chem. Soc.* 91, 1969, p. 5384, copyright (1969) American Chemical Society.)

temperature the lifetimes of the fluorines in the axial and equatorial positions are much greater than the interaction time for the observation of the spectrum, and so they give separate resonances in the spectrum. With increasing temperature, however, the lifetimes of the axial and equatorial fluorines decrease as intramolecular exchange (pseudorotation) accelerates. But the interaction time for the experiment remains unchanged. Eventually, at a sufficiently high temperature, the lifetimes of the equatorial and axial fluorine atoms become much shorter than the interaction time so that the nonequivalent fluorines are no longer distinguished. The spectrum becomes correspondingly simpler. It might therefore be incorrectly concluded that the molecule had a more symmetrical shape, such as a square pyramid, at this temperature. However, if the structure of $(CH_3)_2NPF_4$ is studied by electron diffraction, nonequivalent axial and equatorial fluorines are observed at temperatures at which they are not distinguished by the NMR experiment and at even higher temperatures, because the interaction time for the electron diffraction experiment is much shorter than that for the NMR experiment.

EQUILIBRIUM AND AVERAGE GEOMETRIES

The most complete and unambiguous description of the geometry of a molecule would be the equilibrium structure that is characterized by the r_e equilibrium distance. An equilibrium internuclear distance is the distance between two hypothetically motionless nuclei in a free molecule and corresponds to the minimum of the potential energy function (Figure 1.19). We use the word hypothetical because there is no motionless molecule in reality. Even at the lowest possible temperature (0 K), all molecules possess a certain amount of energy, the zero-point energy of the ground vibrational state, and therefore all the atoms have some motion. However, the equilibrium geometry is a good reference structure, particularly because all theoretical computations yield the equilibrium geometry or information that refers to this geometry.

All experimental data, however, reflect the consequences of averaging over all the intramolecular and intermolecular motions of a molecule. Thus all experimentally determined geometries are average rather than equilibrium geometries. A detailed discussion of the differences between equilibrium and average geometries is beyond the scope of this book, and so we discuss only the following two important questions: (1) What parameters are used to describe average geometries, and (2) which parameter is given by a particular experimental technique?

The following distance parameters may be defined:

r_e: the equilibrium distance between the positions of atomic nuclei corresponding to the minimum of the potential energy

r_α: the distance between average nuclear positions at thermal equilibrium at the temperature T

$r_\alpha^0 (r_z)$:   the distance between average nuclear positions in the ground vibrational state at 0 K

r_g: the average internuclear distance at thermal equilibrium at the temperature T

The difference between r_α and r_α^0 is merely a result of the temperature difference. However r_α, the distance between average nuclear positions, and r_g, the average internuclear distance, are not the same, and they may be very different for a flexible molecule performing large-amplitude vibrations, as illustrated in Figure 1.27. The difference between r_g and r_α may well exceed the experimental error in some cases. Provided that all the necessary corrections have been carefully made, the results of X-ray crystallography, as well as neutron crystallography, microwave spectroscopy, and nuclear magnetic resonance spectroscopy, are r_α parameters. The results of electron diffraction are r_g parameters, but they may be converted to r_α parameters.

Whether or not we need to worry about the differences between these parameters depends on the purpose for which we are using the structural information. For most of our discussions in this book, we will seldom need to specify which distance parameter we are using. Nevertheless, it is useful to be aware of the differences between the parameters, which may be of importance in some special cases. For example, the symmetrical dihalides of the transition metals have a low-frequency bending vibration that has a large amplitude, particularly at the high temperature at which their structures were determined. In such a case the average internuclear distance may be very different from the equilibrium distance or even from the distances between average nuclear positions. If the bond angle is calculated from the r_g distances, an angle of approximately 160° is obtained for all the transition metal dihalides that have been studied to date. But if the r_g parameters are converted to r_α parameters, the calculated angle is 180°, corresponding to a linear molecule. Figure 1.27 shows that the X . . . X distance is smaller during the bending vibration than in the equilibrium structure, which is very close to the structure corresponding to the average nuclear positions. In this and similar cases, a knowledge of the physical meaning of the geometrical parameters is crucial to a proper interpretation of the experimental data.

COMPUTATIONAL TECHNIQUES

The computational determination of molecular geometry has developed considerably in recent years as the capabilities of computers and the theoretical understanding of molecular structure have increased. It is possible, at least in principle, to compute all the properties of a molecule from a knowledge of its composition and connectivity by solving the Schrödinger equation for the molecule. When this is done without invoking any additional information, the technique is called *ab initio*, the Latin expression meaning "from the beginning." If the calculations make use of additional information in the form of parameters experimentally determined for similar molecules, they are called *semiempirical*. Both the ab initio and semiempirical methods generally use approximations in their computations in order to make the often enormous computational task feasible. The main problem with the semiempirical methods is that it is difficult to assess their reliability, especially for molecules that lie outside the range of molecules for which the empirical parameters incorporated in the procedure were derived. Ab initio and semiempirical calculations determine the electronic structure and energy of a molecule for a given arrangement of the nuclei and the positions of the nuclei are adjusted until the configuration of lowest energy is found. There are still severe limitations to the size of the molecule for which ab initio calculations are feasible, but they have no other limitations. They can be used, for example, for molecules that are not amenable to experimental studies, or even for molecules that have never been observed but that may be of theoretical interest. An important feature of ab initio calculations, and in principle of all theoretical calculations, is that they always yield the r_e equilibrium geometry.

There are also several purely empirical techniques that use, for example, molecular force fields based on a large body of experimental data, and it is assumed that these force fields are transferable to other systems. One of the most successful of these methods is molecular mechanics, which expresses the potential energy of the molecule by utilizing geometrical parameters and force constants from experimental data on many other molecules and finds the structure that has the minimum energy.

SOURCES OF DATA

A large amount of experimental data is discussed in this book, especially in Chapters 4, 5, and 6. Wherever possible we have given information on isolated molecules from gas phase studies. But inevitably, much of the information comes from studies on crystals. The best and most complete source of such data is the Cambridge Structural Database (England). The vapor phase structural data were obtained mainly from the Ulm Center for Structure Documentation (Germany).

All parameters are given in SI units. Uncertainties are indicated in parentheses and are given in units of the last digit of the parameter. They have been given whenever they were available, although they have not always been defined in the same way. If the uncertainty is not given, it is because it was not given in the original source, or because the parameters were obtained by ab initio or other calculations for which it has not been customary to quote uncertainties, or because we have calculated the parameter from other experimental data. As a general rule we do not attempt to discuss and interpret differences in structural parameters smaller than 1° in bond angles and 1 pm in bond lengths. We have therefore generally increased stated uncertainties to 0.1° and 0.1 pm.

REFERENCES AND SUGGESTED READING

A. Domenicano and I. Hargittai, Eds., *Accurate Molecular Structures*, Oxford University Press, Oxford, England, 1991.

E. A. V. Ebsworth, D. W. H. Rankin, and S. Cradock, *Structural Methods in Inorganic Chemistry*, Blackwell Scientific Publications, Oxford, England, 1987.

P. J. Wheatley, *The Determination of Molecular Structure*, 2nd Ed., Clarendon Press, Oxford, England, 1968.

_3_

The Valence-shell Electron-pair Repulsion Model

According to the VSEPR model of molecular geometry, the arrangement of the covalent bonds around an atom depends on the total number of electron pairs in the valence shell of the atom, including those that are nonbonding or lone pairs. If there are n atoms X attached to a central atom A by single bonds, and there are m nonbonding or lone pairs, then there are a total of $n + m$ electron pairs in the valence shell of A. The shape of the molecule $AX_n E_m$, where E represents a nonbonding or lone pair, depends on the arrangement adopted by the $n + m$ pairs of electrons in the valence shell of A. The basic assumption of the VSEPR model is that the arrangement of a given number of electron pairs in the valence shell of an atom is that which maximizes their distance apart.

We assume for the present that the inner shells of electrons are complete so that the central core of the atom, consisting of the nucleus and the completed inner shells, is spherical. Such a spherically symmetrical core has no influence on the arrangement of the valence-shell electron pairs. The consequences of a nonspherical central core are discussed in Chapters 5 and 6.

THE POINTS-ON-A-SPHERE MODEL

The arrangements of two to six electron pairs in a spherical shell that maximize their distance apart can be deduced by representing each electron pair by a point on the surface of a sphere and maximizing the least distance between any pair of points. The arrangements deduced in this way are shown in Table 3.1 and Figure 3.1. Two points have a linear arrangement, three points an equilateral triangular

TABLE 3.1 ARRANGEMENTS OF TWO TO SIX
ELECTRON PAIRS THAT MAXIMIZE THEIR DISTANCE
APART

Number of Electron Pairs	Arrangement
2	Linear
3	Equilateral triangle
4	Tetrahedron
5	Trigonal bipyramid
6	Octahedron

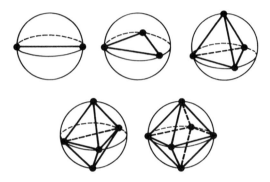

Figure 3.1 Arrangements of points on the surface of a sphere that maximize their distance apart.

arrangement, four points a tetrahedral arrangement, and six points an octahedral arrangement. For five points a square pyramidal arrangement and a trigonal bipyramidal arrangement both maximize the least distance between any pair of points. However, the trigonal bipyramid has only six such least distances of $\sqrt{2}r$, where r is the radius of the sphere, whereas the square pyramid has eight such distances; so the five points are on average farther apart in the trigonal bipyramid arrangement than in the square pyramid arrangement (Figure 3.2).

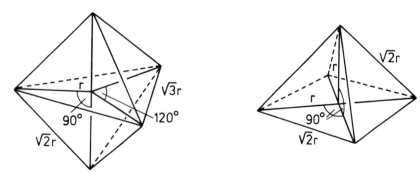

Figure 3.2 The number of least distances, $\sqrt{2}r$, is smaller (6) in a trigonal bipyramid than in a square pyramid (8).

MOLECULAR SHAPES BASED ON TWO TO SIX VALENCE-SHELL ELECTRON PAIRS

The arrangements of three to six electron pairs in Table 3.1 can each give rise to two or more molecular shapes, depending on how many of the electron pairs are nonbonding pairs. These different molecular shapes are summarized in Table 3.2 and Figure 3.3.

TABLE 3.2 ELECTRON-PAIR ARRANGEMENTS AND THE GEOMETRY OF AX_nE_m MOLECULES

Number of Electron Pairs	Arrangement of Electron Pairs	n	m	Class of Molecule	Shape of Molecule	Examples
2	Linear	2	0	AX_2	Linear	BeH_2, $BeCl_2$
3	Equilateral triangular	3	3	AX_3	Equilateral triangular	BCl_3, $AlCl_3$
		2	1	AX_2E	Angular	$SnCl_2$
4	Tetrahedral	4	0	AX_4	Tetrahedral	CH_4, $SiCl_4$
		3	1	AX_3E	Triangular pyramidal	NH_3, PCl_3
		2	2	AX_2E_2	Angular	H_2O, SCl_2
5	Trigonal bipyramidal	5	0	AX_5	Trigonal bipyramidal	PCl_5, AsF_5
		4	1	AX_4E	Disphenoidal	SF_4
		3	2	AX_3E_2	T-shaped	ClF_3
		2	3	AX_2E_3	Linear	XeF_2
6	Octahedral	6	0	AX_6	Octahedral	SF_6
		5	1	AX_5E	Square pyramidal	BrF_5
		4	2	AX_4E_2	Square planar	XeF_4

n = number of bonding pairs, m = number of nonbonding pairs.

Two electron pairs have a linear arrangement, and if both are bonding pairs, they give a linear AX_2 molecule, such as beryllium dichloride, $BeCl_2$, in the gas phase. An AXE molecule is diatomic and is necessarily linear.

Three electron pairs have an equilateral triangular arrangement, so an AX_3 molecule such as boron trichloride, BCl_3, in which all three pairs are bonding pairs has an equilateral triangular shape. An AX_2E molecule such as tin(II) chloride, $SnCl_2$, has a bent or angular shape.

Four electron pairs have a tetrahedral arrangement. Therefore, an AX_4 molecule such as methane, CH_4, has a tetrahedral shape. An AX_3E molecule such as ammonia, NH_3, has a trigonal pyramidal shape, and an AX_2E_2 molecule such as water, H_2O, has an angular shape.

Five electron pairs have a trigonal bipyramidal arrangement, so an AX_5 molecule such as PCl_5 has a trigonal bipyramidal shape. Because the trigonal bipyramid has nonequivalent equatorial and axial vertices, there are alternative positions for nonbonding pairs. For reasons that are discussed later in this

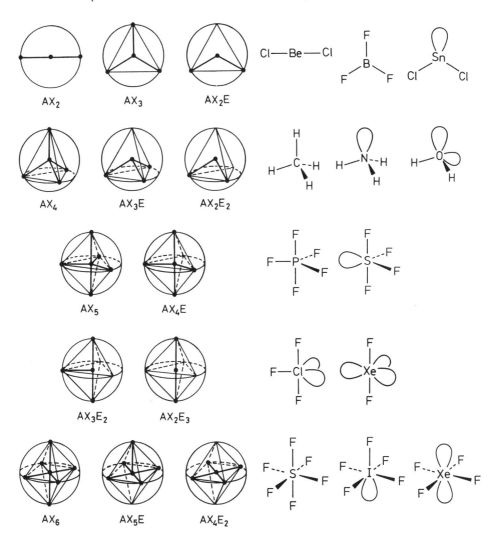

Figure 3.3 Molecular shapes based on the arrangements of two to six valence-shell electron pairs.

chapter, the lone pairs always occupy the equatorial positions; so an AX_4E molecule such as sulfur tetrafluoride, SF_4, has a disphenoidal or seesaw shape. In an AX_3E_2 molecule such as chlorine trifluoride, ClF_3, both lone pairs are in equatorial positions, so the molecule has a T-shaped structure in which all four atoms are in the same plane. In an AX_2E_3 molecule the three lone pairs are all in the equatorial positions, resulting in a linear molecule such as XeF_2.

Six electron pairs have an octahedral arrangement, so an AX_6 molecule such as SF_6 has an octahedral shape. An AX_5E molecule such as IF_5 has a square pyramidal shape. In an AX_4E_2 molecule the two lone pairs may be either *cis* or *trans* to each other. For reasons that are discussed later in the chapter, the two lone pairs always occupy *trans* positions, giving a square planar molecule, such as XeF_4. There are no known AX_3E_3 molecules, but they may be predicted to have a T-shaped structure.

ELECTRON-PAIR DOMAINS

Although the points-on-a-sphere model is useful for predicting the arrangements of a given number of electron pairs in a valence shell, it is not a very realistic model. Because electrons are not stationary and their motion cannot be precisely defined, it is more realistic to think of an electron pair in the valence shell of an atom as a charge cloud that occupies a certain region of space. As a rough approximation, we may think of each electron pair in the valence shell as an impenetrable charge cloud that occupies a certain region of space and excludes other electrons from this space. That electrons behave in this way is in accord with the *Pauli exclusion principle*, which states that the total wave function for any system with two or more electrons must be antisymmetric to electron interchange. As is discussed in more detail in Chapter 7, an important consequence of the Pauli principle is that electrons with the same spin tend to keep apart in space, while electrons of opposite spin may occupy the same region of space. Thus each charge cloud consists of a pair of electrons of opposite spin. This is analogous to the usual statement of the Pauli principle in terms of the atomic orbital model: an orbital can only contain two electrons, which must be of opposite spin. We will call the space occupied by the charge cloud of a pair of electrons of opposite spin an *electron-pair domain*. The relationship between electron-pair domains and orbitals will be discussed in Chapter 7. We will also see in Chapter 7 that it is only a crude approximation to assume that the electron-pair domains in a valence shell are impenetrable and nonoverlapping; nevertheless, it is a very useful model that enables us to make many predictions about the detailed shapes of molecules.

THE SPHERICAL ELECTRON-PAIR DOMAIN MODEL

It was first suggested by G. E. Kimball that an electron-pair domain can be approximately represented by a sphere. H. A. Bent has made considerable use of this idea and he has called it the tangent sphere model.

A useful demonstration model can be constructed by using Styrofoam spheres to represent spherical electron-pair domains. The spheres can be joined in pairs and triples by elastic bands, as shown in Figure 3.4. The elastic bands represent the force of attraction of the central positive core for the valence-shell electron pairs. The point at which the elastic bands are knotted or twisted together represents the central core of the atom. Clusters of two to six electron pairs made in this way automatically adopt the arrangements predicted by the points-on-a-sphere model, as shown in Figure 3.5. These models also show that these are the arrangements that allow a given number of spherical electron-pair

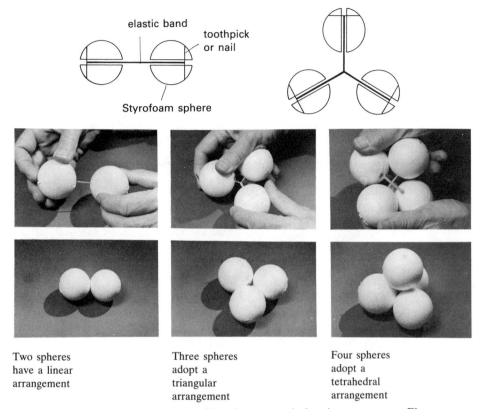

Two spheres
have a linear
arrangement

Three spheres
adopt a
triangular
arrangement

Four spheres
adopt a
tetrahedral
arrangement

Figure 3.4 Construction of models of electron-pair domain arrangements. Electron pair arrangements can be demonstrated by joining two or three Styrofoam spheres by elastic bands threaded through holes in the spheres and held in place with small nails or toothpicks. Each sphere represents a pair of electrons or more exactly the domain of each pair of electrons. The elastic band provides a force of attraction between the spheres that corresponds to the electrostatic attraction between the electrons and a positive core situated at the midpoint of the elastic band. Arrangements of four or more spheres can be demonstrated by twisting together the appropriate number of pairs or triples. The spheres naturally adopt the arrangements shown. If they are forced into some other arrangement, such as the square planar arrangement for four spheres, they immediately adopt the preferred arrangement when the restraining force is removed.

domains to get as close as possible to the central core, thereby minimizing the energy. If any of these models is forced into another less stable arrangement, it will, on gentle shaking, automatically adopt the most stable arrangement. For example, to force four spheres into a square arrangement, their distance from the central core must be increased and the energy of the system correspondingly increased (Figure 3.4). But on gently shaking the model the spheres return to the more stable tetrahedral arrangement.

Another useful demonstration model can be made by tying the appropriate number of balloons together (Figure 3.6). Because balloons are readily obtainable in various shapes and sizes, this model is particularly useful for showing the

Figure 3.5 Styrofoam sphere models representing the arrangements of two, three, four, five, and six-valence shell electron-pair domains.

triangular
arrangement

tetrahedral
arrangement

trigonal
bipyramidal
arrangement

octahedral
arrangement

Figure 3.6 Balloon models representing the arrangements of two to six valence-shell electron-pair domains.

effects of electron-pair domains of different sizes and shapes on molecular shape, as will be discussed later in this chapter.

An interesting analogy was noticed by G. Niac and C. Florea, who pointed out that clusters of walnuts grow with the same arrangements as predicted for electron-pair domains (Figure 3.7).

Figure 3.7 Walnut clusters show the arrangements adopted by electron-pair domains in a valence shell.

Although it is only a rough approximation to assume that electron-pair domains are nonoverlapping, we will see in Chapter 7 that the overlapping of electron-pair domains in a valence shell is minimized in the most stable arrangement in which the electron pairs are as far apart as possible. In any other arrangement, there is more overlapping of the electron-pair domains and the energy of the system is increased. Thus electron pairs behave as if they repel each other and this is the reason for the name VSEPR (valence-shell electron-pair repulsion) model. We will see that in discussing molecular geometry it will generally be more convenient to emphasize the space-occupying properties of electron pairs rather than their mutual repulsion. Moreover, the original emphasis on electron-pair repulsion led to the erroneous idea, sometimes found in discussions of the VSEPR model, that it is a classical electrostatic model and therefore not in accord with the quantum mechanical description of a molecule. The discussion in this book is therefore based mainly on the effect of the different sizes and shapes of electron-pair domains on molecular geometry. An alternative name for the model would be the VSEPD (valence-shell electron-pair domain) model, but we will continue to use the name VSEPR because it is now so well established.

The "spheres and elastic bands" model described above gives the arrangement of a given number of equal spheres in which they are packed as closely as possible around a central point. An alternative way to obtain these same arrangements is to consider the packing of a given number of equal circles, or circular domes, on the surface of a sphere so that they occupy as much as possible of the surface of the sphere. The central points of each circular dome then have the same arrangement as obtained by the points-on-a-sphere model (Figure 3.8).

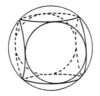

Figure 3.8 The octahedral arrangement of six equal circles on a sphere maximizes the area covered by the circles.

DEVIATIONS FROM IDEAL BOND ANGLES

We have seen that we can predict the general shapes of molecules from the total number of electron pairs in the valence shell of the central atom. However, the bond angles in many molecules are not exactly equal to the ideal angles corresponding to these shapes. Qualitative predictions of these deviations from the ideal bond angles can be made by taking into account the differences in the sizes and shapes of the electron-pair domains in a valence shell. The electron-pair domains in a valence shell are not all equivalent for three important reasons:

1. Nonbonding or lone pairs have larger domains than bonding pairs.
2. Bonding domains decrease in size in the valence shell of the central atom A with increasing electronegativity of the ligand X and increase in size with increasing electronegativity of the central atom.
3. Double-bond and triple-bond domains that consist of two and three electron pairs, respectively, are larger than single-bond domains.

Nonbonding or Lone Pairs

Because a nonbonding pair is subject only to the attraction of one positively charged core, it is pulled in toward the core and tends to spread out and surround the core as far as is permitted by the presence of the other valence-shell electron pairs. A bonding pair, however, is subject to the attraction of two positive cores, and therefore it occupies a smaller and more contracted domain that is partly in the valence shell of A and partly in the valence shell of the ligand X. Thus a bonding pair has a domain that occupies a smaller volume of the valence shell of the central atom A than does a nonbonding pair and is farther from the core of the central atom than a nonbonding pair. The same idea was expressed in the original formulation of the VSEPR model in terms of electron-pair repulsions by the statement that the magnitude of electron-pair repulsions decreases in the order

lone pair : lone pair > lone pair : bond pair > bond pair : bond pair

Because a nonbonding domain occupies a larger fraction of the valence shell than a single-bond electron-pair domain, the angles between the three bond domains in an AX_3E molecule are smaller than the angles between the bond domains and the lone-pair domain. We will assume for the present that the angles between the bonds are the same as between the charge centroids of the bond domains. There is good reason to suppose that this is usually true, at least to a very good approximation. We will consider later a few cases where this might not be a valid assumption. Thus in an AX_3E molecule the bond angle is expected to be smaller than the ideal angle of 109.5° (Figure 3.9).

Alternatively, because lone pair:bond pair repulsions are greater than bond pair:bond pair repulsions, and because there are three lp:bp and three bp:bp repulsions, the angles between the lone pair and the bond pairs are predicted to be greater than the angles between the bond pairs. Therefore, it is expected that the XAX bond angle in an AX_3E molecule will be smaller than the ideal tetrahedral angle (Figure 3.9). In general, in the determination of the geometry of a molecule it is the positions of the nuclei that are found, and the lone pairs are not directly located, although the position of one lone pair can generally be inferred by symmetry. For example, it is reasonable to assume that the charge centroid of the lone pair is on the threefold axis of an AX_3E molecule. We can then find the angle XAE (in degrees) from the relationship

$$\cos (XAE - 90) = [\tfrac{2}{3}(1 - \cos XAX)]^{1/2}$$

In PF_3, for example, the experimental bond angle is 96.9° and the FPE angle is then calculated to be 120.2°.

Because the two lone-pair domains in an AX_2E_2 molecule occupy a greater total volume of the valence shell than a single lone pair, we might expect that the bond angle would be smaller than in a related AX_3E molecule. The bond angle in H_2O, for example, is 104.5°, which is less than the bond angle of 107.5° in the NH_3 molecule. This is not always the case, however. For example, the bond angle in NF_3 is 102.1°, but that in OF_2 is 103.2°. However, we must be careful with such comparisons. Oxygen is more electronegative than nitrogen, so the OF bonds are less polar than the NF bonds and they therefore repel each other more strongly than the NF bonds. This is probably the reason why the FOF angle is larger than the FNF angle. But, in any case, it is not possible to predict with certainty whether the bond angle in an AX_2E_2 molecule will be larger or smaller than that in a related AX_3E molecule. The lone pairs are in a plane perpendicular to the plane of the bond pairs, and so the effect that they might have on the bond angle

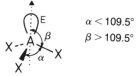

$\alpha < 109.5°$
$\beta > 109.5°$

Figure 3.9 The bond angles in an AX_3E molecule are smaller than the ideal tetrahedral angle.

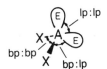

Figure 3.10 In an AX_2E_2 molecule there are one lp:lp interaction, one bp:bp interaction, and four bp:lp interactions.

is difficult to predict. We might assume, for example, that the resultant effect of the two lone pairs along the bisector of the XAX angle would be $2 \cos (109.5/2)$ = 1.2 times the effect of one lone pair in this position. But we would then be dealing with an effective AX_2E molecule in which the ideal bond angle is 120°, and so it is not possible to predict if the effect of the larger lone-pair domains would be to decrease the angle to less than 109.5°. Alternatively, we may note that there is only one lp:lp repulsion and only one bp:bp repulsion, but four lp:bp repulsions (Figure 3.10); and since we do not know their relative magnitudes, it is not possible to make any firm prediction about the bond angle. In many AX_2E_2 molecules the bond angle is smaller than the ideal tetrahedral angle, as we see from the examples in Table 3.3.

TABLE 3.3 BOND ANGLES IN SOME AX_2E_2 MOLECULES

Molecule	Bond Angle (°)
H_2O	104.5(1)
F_2O	103.1(1)
SCl_2	103.0(4)
SF_2	98.0(1)
$S(CF_3)_2$	97.3(8)
$Se(CF_3)_2$	96(2)
$Te(CF_3)_2$	90(1)

The presence of one or more lone pairs in a valence shell also affects bond lengths. Because a lone pair is attracted more closely to the central core and has a larger domain than the neighboring bond pairs, it prevents the neighboring bond pairs from getting as close to the central core as they would in its absence; so the bonds adjacent to a lone pair are longer than they would be in the absence of the lone pair. In AX_3E and AX_2E_2 molecules the lone pairs interact equally with all the bonds, so the effect on the bond lengths cannot be observed. However, in AX_4E, AX_3E_2, and AX_5E molecules the bonds that are adjacent to the lone pair are always longer than the bonds that are farther removed from the lone pair. For example, in BrF_5 the bonds in the base of the square pyramid have a length of 179 pm, whereas the axial bond has a length of 168 pm.

In an AX_4E_2 molecule in which there is an octahedral arrangement of electron-pair domains, there are two alternative positions for the two lone-pair domains; they may be either *cis* or *trans* to each other. To minimize their overlap, the lone-pair domains always occupy the *trans* positions, giving rise to square planar molecules such as XeF_4 and ICl_4^-.

Effect of Ligand Electronegativity

An electronegative ligand draws the bonding charge density away from the central atom so that the bonding-pair domain is unsymmetrical and occupies more space in the valence shell of the ligand than it does on the central atom. With increasing electronegativity of the ligand, the space occupied by the bonding electron-pair domain in the valence shell of the central atom decreases (Figure 3.11). Consequently, bond angles between bonds to more electronegative ligands are smaller than those to bonds to less electronegative ligands. Some examples are given in Table 3.4.

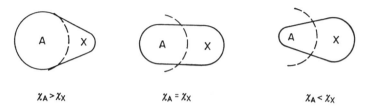

$$\chi_A > \chi_X \qquad\qquad \chi_A = \chi_X \qquad\qquad \chi_A < \chi_X$$

Figure 3.11   The size of the bonding electron-pair domain in the valence shell of A decreases with increasing electronegativity of the ligand X.

TABLE 3.4   VARIATION OF BOND ANGLES WITH LIGAND ELECTRONEGATIVITY

PX_3	XPX (°)	OSX_2	XSX (°)
PF_3	97.8(2)	OSF_2	92.3(3)
PCl_3	100.3(9)	$OSCl_2$	96.2(7)
PBr_3	101.0(4)	$OSBr_2$	98.2(2)

Multiple Bonds

A double bond consists of two pairs of shared electrons and a triple bond consists of three pairs of shared electrons. The general arrangements of a given number of electron pairs shown in Table 3.1 and Figure 3.1 are independent of whether they are bonding or nonbonding pairs or whether they form single, double, or triple bonds. In ethene there are four pairs of electrons in the valence shell of each carbon atom, and these four pairs have the expected tetrahedral arrangement as in ethane. Figure 3.12 shows that if two of these pairs are shared between the two carbon atoms forming a double bond, the ethene molecule has an overall planar shape. According to this model, the double bond consists of two shared pairs, one

Figure 3.12   Overlap of two electron-pair domains to give a four-electron double-bond domain in the ethene molecule. There is a tetrahedral arrangement of four electron-pair domains around each carbon atom. There is a planar triangular arrangement of two single-bond and one double-bond domains around each carbon atom.

on each side of the molecular plane. The carbon-carbon bonds in this model of ethene are often described as bent bonds because the electron-pair domains are not located on the internuclear axis but one on each side of the axis. Figure 3.13 shows that if three electron pairs are shared between two carbon atoms to form a triple bond, as in the ethyne molecule, the molecule has a linear shape. The three electron-pair domains are clustered around the internuclear axis in a triangular arrangement, and the triple bond can be described as consisting of three bent bonds.

Although the bent-bond model of double and triple bonds is a valid and useful description, we should beware of interpreting it too literally. Because electron-pair domains are not as localized as our model assumes and, in particular, because they overlap each other, there are not in fact two separate regions of electron density on either side of the axis of a double bond. The two domains overlap to form a two-electron-pair double-bond domain in which the two separate pairs cannot be distinguished and that has its maximum charge density on the internuclear axis (Figure 3.12). Unlike a single-bond domain, a double-bond domain is not symmetrical around the internuclear axis but can be approximately described as having a prolate ellipsoidal shape with its long axis perpendicular to the plane of the ethene molecule (Figure 3.12). Similarly, the three spherical domains of a triple bond overlap to form a single three-electron-pair domain that again has its maximum charge density on the internuclear axis and can be described as having an oblate ellipsoidal shape with its short axis along the internuclear axis (Figure 3.12). The alternative but equivalent model of double and triple bonds that describes them as consisting of σ and π orbitals is described in Chapter 7.

Figure 3.13 Overlap of three electron-pair domains to give a six-electron pair triple-bond domain. There is a tetrahedral arrangement of four electron-pair domains around each carbon atom. There is a linear arrangement of a single-bond domain and a triple-bond domain around each carbon atom.

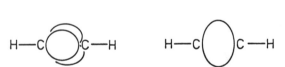

We can use the concept of double- and triple-bond domains as the basis for a very simple method for predicting the shapes of molecules containing double and triple bonds that is particularly convenient when there are more than four electron pairs in the valence shell of the atom forming a double or triple bond. Instead of considering the arrangement of a given number of electron pairs in the valence shell of an atom, we consider the arrangement of the total number of domains that may be lone-pair, single-bond, double-bond, or triple-bond domains. We can then very simply predict the general shape of any molecule containing double and triple bonds, as shown in Figure 3.14. According to this model, the arrangement of the bonds around an atom depends only on the total number of domains (lone pair, single bond, double bond, or triple bond) in its

Total Number of Bonds and Lone-Pairs	Arrangement	Number of Bonds	Number of Lone Pairs	Shape of Molecule	Examples
2	Linear	2	0	Linear	$O=C=O$ $\qquad$ $H-C\equiv N$
3	Triangular	3	0	Triangular	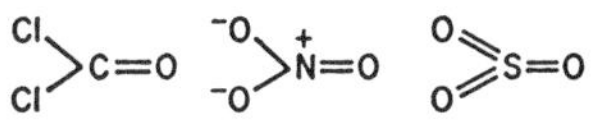 $Cl_2C=O$ $\qquad$ NO_2^+ $\qquad$ SO_2
		2	1	V-Shape	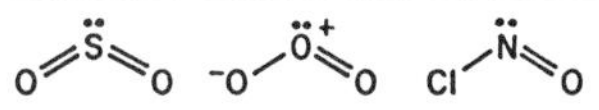 SO_2 $\qquad$ O_2 $\qquad$ $ClNO$
4	Tetrahedral	4	0	Tetrahedral	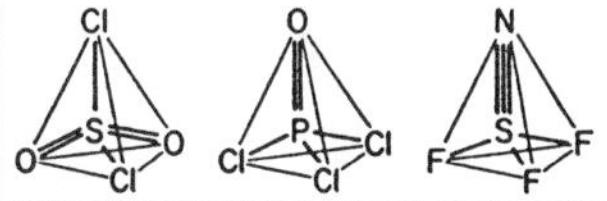 SO_2Cl_2 $\qquad$ $POCl_3$ $\qquad$ SO_2F_2
		3	1	Trigonal Pyramid	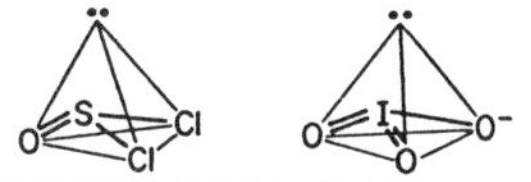 $SOCl_2$ $\qquad$ IO_3^-
		2	2	V-Shape	 ClO_2^- $\qquad$ XeO_2
5	Trigonal Bipyramid	5	0	Trigonal Bipyramid	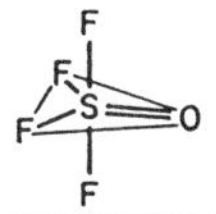 SOF_4
		4	1	Irregular Tetrahedron	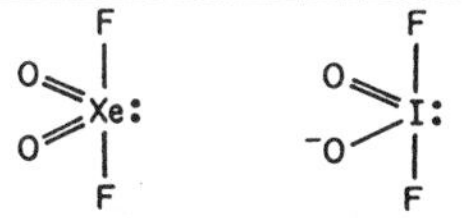 XeO_2F_2 $\qquad$ $IO_2F_2^-$
6	Octahedron	6	0	Octahedron	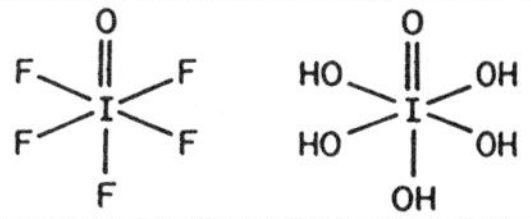 $IO_2F_4^-$ $\qquad$ $I(OH)_4O_2$

Figure 3.14 Geometry of molecules containing double and triple bonds.

valence shell and is independent of the nature of the bonds, that is, whether they are single bonds, double bonds, or triple bonds or have an intermediate character.

The bond angles will, however, depend on the nature of the bonds, because a double-bond domain is larger than a single-bond domain, and a triple-bond domain is larger still. Thus the angles made by triple bonds will be larger than those made by double bonds, which in turn will be larger than those made by single bonds. The structure of dimethylsulfate $(CH_3O)_2SO_2$ provides a good example as it has three different kinds of SO_2 bond angles that decrease in magnitude in the expected order:

$$O=S=O > O=S-O > O-S-O$$

as we see in Figure 3.15.

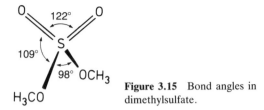

Figure 3.15   Bond angles in dimethylsulfate.

Multicenter Bonds

In some molecules a single electron pair bonds not just two nuclei as in a normal single bond but three or more nuclei. The electron pair is said to be forming a three- or four-center bond or in general a multicenter bond (Figure 3.16).

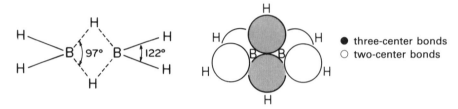

● three-center bonds
○ two-center bonds

Figure 3.16   Three-center bonds in the diborane molecule.

Molecules containing multicenter bonds are called electron deficient because they do not possess enough electrons to join all the bonded atoms by single electron-pair bonds. Diborane, B_2H_6, is such a molecule. Diborane has two three-center bonds, the B---H---B bridge bonds, and four ordinary single electron-pair B—H bonds. Thus each boron has four electron-pair domains in its valence shell, two single-bond domains, and two three-center bond domains. These four domains have the expected tetrahedral arrangement and they determine the shape of the diborane molecule (Figure 3.16). This shape is closely related to that of ethene. Both molecules have exactly the same arrangement of electron-pair domains, but the two electron-pair domains that form the double bond in ethene are three-center bond domains in diborane.

Because a multicenter bond domain is shared between three or more valence shells, it occupies less space in a given valence shell than a single-bond domain. Whereas multiple bonds have bond orders greater than 1, a multicenter bond has a bond order of less than 1. Thus, whereas the angles formed by multiple bonds are larger than those formed by single bonds, the angles formed by multicenter bonds are smaller than those formed by the single bonds in the same valence shell. In diborane, for example, there are three different kinds of HBH bond angles, and they decrease in size in the order

$$H—B—H > H—B---H > H---B---H$$

122°   122°   97°

TRIGONAL BIPYRAMIDAL MOLECULES

Molecules in which there are five domains with a trigonal bipyramidal arrangement in the valence shell of the central atom exhibit a number of features of special interest that merit a separate discussion. We will see that they also exemplify all the consequences of the differences between lone-pair, single-bond, and multiple-bond domains.

The tetrahedron and octahedron are regular polyhedra in which all the vertices are equivalent; each has the same number of nearest neighbors in the same directions and at the same distance. But in the trigonal bipyramid the five vertices are not all equivalent. The apical vertices have three nearest neighbors at 90°, whereas the equatorial vertices have only two nearest neighbors at 90° and two more at 120° (Figure 3.17). This difference has some important consequences

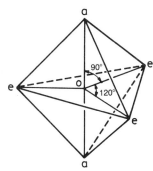

Figure 3.17 Nonequivalence of the axial and equatorial positions in a trigonal bipyramid.

for the shapes of molecules based on the arrangement of five domains in a valence shell. In tetrahedral AX_4 or octahedral AX_6 molecules the bonding electron-pair domains are all equivalent in every way and in the most stable arrangement are all at the same distance from the core, so the bonds are all the same length. But in a trigonal bipyramid, AX_5, molecule the five bonding electron-pair domains are not all equivalent. In particular, the domains in the axial positions have more close neighbors than the domains in the equatorial positions. In other words, the axial positions are more crowded than the equatorial positions. Thus, to minimize overlap with their neighbors, the axial bond-pair domains are farther from the central core than the equatorial bond domains in the equilibrium structure. Thus the axial bonds are predicted to be longer than the equatorial bonds, as is indeed found for trigonal bipyramidal AX_5 molecules, where A is a main group element (Table 3.5). We may say that there is less space available for an electron-pair domain in an axial position of a trigonal bipyramid than in an equatorial position, so an electron-pair domain in an axial position cannot get as close to the central core as an electron-pair domain in an equatorial position, and therefore the axial bonds are longer than the equatorial bonds.

That the equatorial positions are less crowded than the axial positions of a trigonal bipyramid can be demonstrated with the "spheres and elastic bands" model. In Figure 3.18 we see that if the five spheres are all at the same distance

TABLE 3.5 EQUATORIAL AND AXIAL RADII FOR AX_5, AX_4E, AND AX_3E_2 MOLECULES[a]

	r_{eq} (pm)	r_{ax} (pm)	r_{ax}/r_{eq}
AX_5			
PCl_5	105	120	1.14
PF_5	89	94	1.06
$P(C_6H_5)_5$	108	122	1.13
CH_3PF_4	90	97	1.08
$(CH_3)_2PF_3$	91	100	1.10
Cl_2PF_3	95	106	1.12
$SbCl_5$	132	144	1.09
$(C_2H_2Cl)_3SbCl_2$	138	146	1.06
$(CH_3)_3SbCl_2$	130	150	1.15
$(CH_3)_3SbBr_2$	130	149	1.15
$(CH_3)_3SbI_2$	130	155	1.19
$(C_6H_5)_3BiCl_2$	147	161	1.10
$(CH_3)_2SnCl_3^-$	142, 136	155	1.11
AX_4E			
SF_4	91	101	1.11
OSF_4	90	96	1.07
$(C_6H_5)_2SeBr_2$	114	136	1.19
$(C_6H_5)_2SeCl_2$	114	131	1.15
$(CH_3C_6H_4)_2SeCl_2$	117	140	1.20
$(CH_3C_6H_4)_2SeBr_2$	117	140	1.20
$(CH_3)_2TeCl_2$	133	152	1.14
$(C_6H_5)_2TeBr_2$	137	154	1.12
$O_2IF_2^-$	127	136	1.07
AX_3E_2			
ClF_3	94	104	1.11
$C_6H_5ICl_2$	123	146	1.19
BrF_3	108	118	1.09

[a]Where it is necessary to distinguish between axial and equatorial ligands, the equatorial ligands are written in the formula before the central atom.

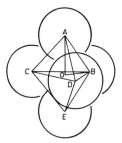

Figure 3.18 Spherical electron-pair domain model of a trigonal bipyramidal arrangement of five valence-shell electron pairs. If $OA = OB$, $BD = \sqrt{2}AB$. If $BD = AB$, $AO = \sqrt{2}OB$.

from the central core the axial spheres are touching all three neighbors, but the equatorial spheres are only touching their two axial neighbors. If the equatorial spheres are allowed to approach the central core more closely, thereby pushing the axial spheres farther away, when the equatorial spheres finally touch and each sphere is then the same distance from all its nearest neighbors, the trigonal bipyramid consists of two tetrahedra sharing a face. In this limiting case the axial spheres are at a distance of $\sqrt{2} = 1.41$ times the distance of the equatorial spheres from the core. We expect therefore that in trigonal bipyramidal AX_5 and related molecules the ratio of the axial to the equatorial covalent radius of the central atom will be between 1.00 and 1.41. This is indeed the case for all known AX_5, AX_4E, and AX_3E_2 molecules of the main group elements. Some examples are given in Table 3.5.

The nonequivalence of the equatorial and axial positions of a trigonal bipyramid has several other important consequences. Because there is more space available in a valence shell in an equatorial position than in an axial position, larger electron-pair domains, in particular those of lone pairs and multiple bonds, preferentially occupy the equatorial positions, as we stated earlier and as is illustrated in Figures 3.19 and 3.20.

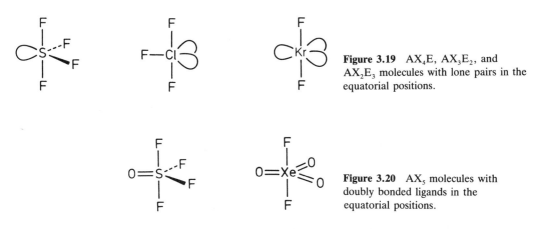

Figure 3.19 AX_4E, AX_3E_2, and AX_2E_3 molecules with lone pairs in the equatorial positions.

Figure 3.20 AX_5 molecules with doubly bonded ligands in the equatorial positions.

In AX_5 molecules with ligands of different electronegativities it is found, as expected, that the less electronegative ligands for which the bond domains occupy more space in the valence shell are always found in the equatorial positions, while the more electronegative ligands are always found in the more crowded axial positions, as is the case for PF_3Cl_2 and PF_2Cl_3, for example (see Chapter 5).

We noted earlier that the least distance between any two electron pairs is maximized for both the square pyramid and trigonal bipyramid arrangements of five electron pairs. But, whereas the square pyramid has eight such least distances, the trigonal bipyramid has only six. So there is somewhat more overlap between the electron-pair domains in a square pyramidal arrangement than in a trigonal bipyramidal arrangement. Hence the trigonal bipyramid has a slightly lower energy than the square pyramid, so AX_5 molecules have a trigonal bipyramidal shape. But because the square pyramidal arrangement has only a slightly higher energy than the trigonal bipyramidal arrangement, an intramolecular exchange of axial for equatorial positions (pseudorotation) is possible in such molecules via a square pyramidal transition state, as we have seen in Chapter 1.

MOLECULAR SHAPES BASED ON VALENCE SHELLS WITH SEVEN, EIGHT, AND NINE ELECTRON PAIRS

Maximizing the least distance between points on a sphere leads to the arrangements for seven, eight, and nine points given in Figure 3.21 and Table 3.6. These are therefore the predicted shapes for AX_7, AX_8, and AX_9 molecules.

monocapped   square   tricapped
octahedron   antiprism   trigonal prism

Figure 3.21 Arrangements of seven, eight, and nine points on a sphere that maximize the least distance between any pair of points.

TALBE 3.6 ARRANGEMENTS OF SEVEN, EIGHT, AND NINE VALENCE-SHELL ELECTRON PAIRS

Number of Electron Pairs	Predicted Geometry	Other Observed Geometries
7	Monocapped octahedron	Monocapped triangular prism Pentagonal bipyramid
8	Square antiprism	Dodecahedron
9	Tricapped triangular prism	—

In addition to the monocapped octahedral shape predicted by the hard-sphere model, molecules with seven electron pairs in the valence shell of the central atom are also found with a pentagonal bipyramidal and a monocapped triangular prism shape (Figure 3.22). The hard-sphere model corresponds to a repulsive force between the points on the sphere that follows a force law of the type $F \propto 1/r^n$, where $n \to \infty$. But if n has a finite value, then it has been shown that the pentagonal bipyramid is the most stable arrangement for values of n up to 3, and then the monocapped triangular prism is the most stable arrangement for $3 < n < 6$, and for $n > 6$ the monocapped octahedron is the most stable arrangement. Unfortunately, because we do not know which force law it is appropriate to use in a given case, no firm prediction of the structure of an AX_7 molecule can be made. Moreover, there is the added complication that in these molecules the observed bond lengths are not all equal, so the points-on-a-sphere model is only an approximation.

For eight points on a sphere the predicted square antiprism arrangement is independent of the value of n in the force law $F \propto 1/r^n$, so no other shape for an AX_8 molecule is expected. Although many AX_8 molecules have the predicted square antiprism shape the triangular dodecahedral (bisdisphenoidal) shape is also common (Figure 3.23). In these dodecahedral molecules the bond lengths are

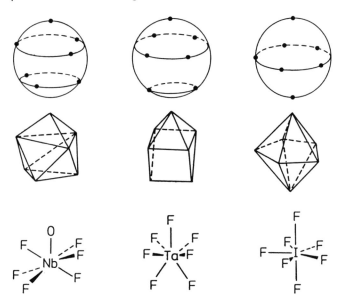

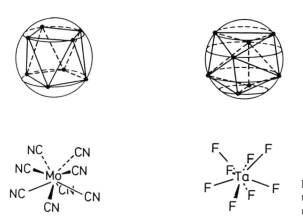

Figure 3.22 Monocapped octahedral, monocapped trigonal prism, and pentagonal bipyramidal AX_7 molecules.

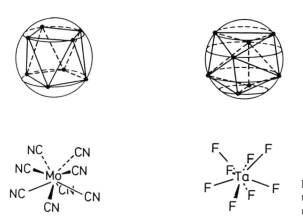

Figure 3.23 Square antiprism and triangular dodecahedral AX_8 molecules.

not all equal, and this may well be the factor that causes the dodecahedral shape to be favored over the square antiprism.

For nine points on a sphere, the predicted tricapped triangular prism arrangement (Figure 3.21) is also independent of the value of n and is the only shape that has been found for AX_9 molecules (Figure 3.24).

The only molecules in which lone pairs are found in a valence shell with more than six electron pairs are those of the AX_6E and AX_5E_2 types. The structures of these molecules are discussed in Chapter 5. It is unlikely that many other examples will be found of molecules in which the central atom has a valence shell of more than six electron pairs including one or more lone pairs, because such large valence shells are not common and are found mostly for the transition

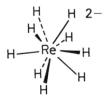

Figure 3.24 Tricapped trigonal prism AX_9 molecule.

elements and the lanthanides and actinides. For these elements any unshared electrons are not in the valence shell but they occupy inner d and f orbitals. These electrons either have no effect on the arrangement of the bonding pairs or they have a smaller and different effect from that of lone pairs in the valence shell of a main group element, as we discuss in Chapter 6.

LIMITATIONS

Although the VSEPR model has very wide applicability, it is important to recognize its limitations, which we summarize here.

1. The VSEPR model assumes that the bonding in a molecule or crystal has an appreciable covalent character. It is therefore not applicable to those molecules and crystals in which the bonding is predominantly ionic.

2. The VSEPR model assumes that the electron pairs in a valence shell are localized bonding and nonbonding pairs. Thus it is primarily concerned with those molecules in which the bonds can be described as two-center, two-electron bonds. It can also be satisfactorily applied to molecules in which the bonds can be described as three-center, two-electron bonds; but if the bonding electrons are still more delocalized, the VSEPR model becomes much less useful. However, if only some of the electron pairs are delocalized, the geometry may be predicted by simply considering the arrangement of the localized pairs and ignoring any extensively delocalized electrons. For example, the planar hexagonal shape of the benzene molecule can be predicted on the basis of the arrangement of the three localized electron pairs in the valence shell of each carbon atom.

3. When the ligand atoms or groups are sufficiently large compared to the central atom, steric interactions between the ligands may prevail over the bond–bond interactions and the VSEPR predictions may no longer be valid.

4. Molecules whose shapes are based on valence shells with seven or eight electron pairs cannot always be unambiguously predicted, because there are two or three electron-pair arrangements that do not differ greatly in stability. In this case, ligand–ligand interactions, the effects of chelating ligands, and other small effects may influence which of these electron-pair arrangements is adopted in any given case.

5. Some chelating ligands have a geometry that may dictate the geometry around the central atom. This is particularly the case for high coordination numbers, where several different geometries may have similar energies.

6. In its simplest form the VSEPR model assumes a spherical core that does not affect the arrangement of the electron pairs in the valence shell. If the spherical symmetry of the core is deformed, this must be taken into account and the predictions of the simple VSEPR model modified accordingly, as we will see in Chapters 5 and 6.

7. The VSEPR model considers free, isolated molecules. Thus it is applicable to gas-phase structures and to those crystal structures in which the intermolecular (or interionic) interactions are sufficiently weak that they may be ignored. In many crystal structures in which there are significant intermolecular interactions, these interactions can often be considered to result from weak bonds between the molecules (or ions). The VSEPR model can then usually be applied if the weak bonds are also taken into consideration, as we will see in Chapter 5.

REFERENCES AND SUGGESTED READING

H. A. BENT, *J. Chem. Ed.*, **40**, 446, 523, 1963; **42**, 302, 348, 1965; **44**, 512, 1967; **45**, 768, 1968.

R. J. GILLESPIE, *J. Chem. Ed.*, **40**, 295, 1963.

R. J. GILLESPIE, *J. Chem. Ed.*, **47**, 18, 1970.

R. J. GILLESPIE, *Molecular Geometry*, Van Nostrand Reinhold, London, 1972.

R. J. GILLESPIE, and R. S. NYHOLM, *Quart. Rev. Chem. Soc.*, **11**, 339, 1957.

I. HARGITTAI and B. L. CHAMBERLAND, "The VSEPR Model of Molecular Geometry," in I. HARGITTAI, Ed., *Symmetry: Unifying Human Understanding*, Pergamon Press, New York, 1986.

G. E. KIMBALL, references to unpublished work by Kimball and his students are given by Bent.

G. NIAC and C. FLOREA, *J. Chem. Educ.*, **57**, 429, 1980.

N. V. SIDGWICK and H. E. POWELL, *Proc. Roy. Soc.*, **A 176**, 153, 1940.

_4_

The Second-period Elements

The electronegativity of the elements of the second period increases from a very low value for lithium (1.0) to a very large value for fluorine (4.1). With the exception of its compounds to other metals in which the bonding is metallic, lithium is almost always bonded to another atom of considerably higher electronegativity with which it forms a predominantly ionic bond. In such ionic compounds there is very little, if any, electron density in the valence shell of the lithium atom. Thus the VSEPR model is not appropriate for discussing the geometry of the compounds of lithium. But the elements from beryllium to fluorine form a large number of predominantly covalent compounds whose geometry can be predicted by means of the VSEPR model. In their covalent compounds these elements are limited to a maximum of four pairs of electrons in their valence shells. The possible geometries of these elements are therefore limited to those summarized in Table 4.1.

As the core charge increases from beryllium to fluorine, the valence shell becomes smaller and the four electron pairs become increasingly crowded. The small size of the valence shell of each of these elements has several important consequences. (1) Deviations from the ideal tetrahedral angle are generally rather small in this period compared to the deviations that are observed in compounds of elements in the third and subsequent periods, in which the central atom has only four pairs of electrons in its considerably larger valence shell. (2) When the ligands are large, in particular when they are from period 3 and beyond, ligand–ligand repulsions are generally important, and they frequently increase the bond angles to values larger than the tetrahedral angle. But when the ligands are other atoms from period 2, ligand–ligand repulsions are small or negligible, and

TABLE 4.1 MOLECULAR GEOMETRIES FOR THE ELEMENTS BERYLLIUM TO FLUORINE

Single Bonds					
Number of electron pairs	2	3	4		
Number of lone pairs	0	0	0	1	2
Shapes	Linear	Trigonal planar	Tetrahedral	Pyramidal	Angular
	$-Be-$	Be^-	Be^{2-}		
	$-B^+-$	B	B^-		
		C^+	C	$\ddot{C}^-$	
			N^+	$\ddot{N}$	$:\ddot{N}^-$
			O^{2+}	$\ddot{O}^+$	$\ddot{O}$
					$\ddot{F}^+$

Multiple Bonds			
Number of domains	2	3	
Number of lone pairs	0	0	1
Shapes	Linear	Trigonal planar	Angular
	$=B-$	B^-	
	$-C\equiv$	C	$\ddot{C}^-$
	$=C=$		
	$-N^+\equiv$	N^+	$\ddot{N}$
	$=N^+=$		
	$-O^{2+}\equiv$	O^{2+}	$\ddot{O}^+$
	$=O^{2+}=$		

they have little effect on the bond angles which are determined by the strong interactions between the valence shell electron pairs. (3) Because the larger valence shells of the third and subsequent period elements are not filled by four pairs of electrons, they can accommodate at least two more electron pairs, as is discussed in Chapter 5. Thus, when nitrogen, oxygen, or fluorine is bonded to a third or subsequent period element that has an incompletely filled valence shell, there is a tendency for the strong repulsions between the electron pairs in the valence shell of these elements to be reduced by the delocalization of at least one of the nonbonding electron pairs into the valence shell of the heavier element. The domain of the "nonbonding" electron pair then occupies less space in the valence shell of nitrogen, oxygen, or fluorine, and its effect on the bond angles is correspondingly reduced.

BERYLLIUM

Beryllium has a strong tendency to complete its valence shell of four electron pairs, so compounds with only two or three electron pairs in the valence shell of beryllium are relatively uncommon and are generally found only in the gas phase or when the ligands are sufficiently large to cause steric crowding around the beryllium atom. In the majority of the compounds of beryllium there are four electron pairs in its valence shell, and it has a formal charge of -2. However, because the bonds are from beryllium to more electronegative elements, these bonds are polar, and the actual charge of beryllium is closer to zero than -2. Since beryllium has only two valence electrons there are no compounds in which beryllium has unshared pairs in its valence shell.

AX_2 Linear Geometry

The halides BeF_2, $BeCl_2$, $BeBr_2$, and BeI_2 have the expected linear structure in the gas phase at high temperature. Dimethylberyllium, $(CH_3)_2Be$, and di-tert-butylberyllium, $[(CH_3)_3C]_2Be$, have linear structures in the gas phase, but dimethylberyllium is polymeric in the solid state (see page 68).

Beryllium forms a series of cyclopentadienyl complexes $Be(C_5H_5)X$, with $X = H$, Cl, Br, CH_3 and $CH=CH_2$, which have the structure shown in Figure 4.1. Although it is not possible to write a completely satisfactory Lewis structure for a molecule of this type, we can give an approximate description by means of which we can predict the geometry. We first imagine the molecule to have the ionic structure $Cp^-Be^+—X$ in which the cyclopentadienyl ion can be represented by the resonance structures in Figure 4.2. We can then imagine that two double-bond electron pairs and the lone pair on one carbon atom are shared with the positively charged beryllium atom to give a kind of triple bond, as shown in Figure 4.2. We can conveniently write this structure as

$$\overset{2+}{Cp}\!\!\equiv\!\!\overset{2-}{Be}—X$$

In this Lewis structure the beryllium atom is forming four bonds and has a formal

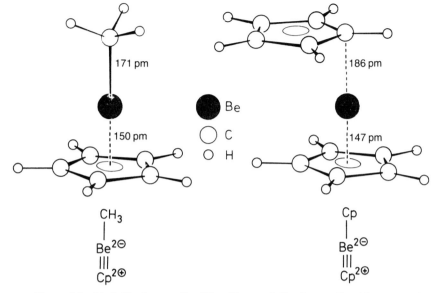

Figure 4.1 Methyl(cyclopentadienyl)beryllium and dicyclopentadienyl beryllium.

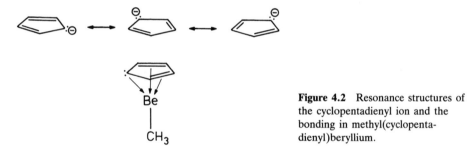

Figure 4.2 Resonance structures of the cyclopentadienyl ion and the bonding in methyl(cyclopentadienyl)beryllium.

charge of -2 as in many of its compounds, and the cyclopentadienyl group has a formal charge of $+2$. In fact, because of the polarity of the bonds the real charges will be smaller. Thus the molecule has a linear AX_2 geometry arising from the linear arrangement of one three-electron-pair domain and one single electron-pair domain in the valence shell of the beryllium atom. The Be—Cl bond in Cl—$Be^{2-}\equiv Cp^{2+}$ has a length of 183.7(6) pm, which is considerably longer than in $BeCl_2$ (175 pm). This increased length of the Be—Cl bond can be attributed to the greater volume of the three-electron-pair domain and the greater repulsion that it exerts compared to the single-electron-pair domain in the Be—Cl bonds in $BeCl_2$.

Dicyclopentadienyl beryllium is also known, but it does not have a symmetrical structure like that of ferrocene (Chapter 6). In the solid state one of the cyclopentadienyl groups is bonded in the same symmetrical manner as in the Cp—Be—X molecules, but the other is bonded through only one carbon atom, as shown in Figure 4.1. Thus we can write the structure of dicyclopentadienyl beryllium as $\overset{2+}{Cp}\!\!\equiv\!\!\overset{2-}{Be}\!\!-\!\!Cp$ in which beryllium forms one triple bond and one single bond. We would not expect dicyclopentadienyl beryllium to have the same symmetrical structure as ferrocene, because beryllium is limited to four pairs of electrons in its valence shell and therefore cannot form triple bonds to two cyclopentadienyl rings.

AX_3 Trigonal Planar Geometry

Beryllium dichloride and beryllium dibromide form dimers in the vapor state that have a four-membered ring structure in which both beryllium atoms have a triangular AX_3 geometry. This geometry is also observed in the dimer of $Be[OC(CF_3)_3]_2$ (Figure 4.3). The approximately 90° bond angles in the four-membered rings of these dimeric molecules are far from the ideal values of 120° at beryllium and 109° at oxygen or chlorine, but presumably the bonds are bent as in most other small rings, so the angles between the electron pairs are much closer to the ideal values.

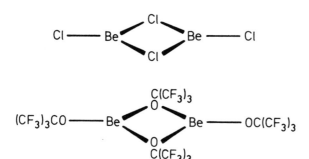

Figure 4.3 Triangular AX_3 geometry in $(BeCl_2)_2$ and $\{Be[OC(CF_3)_3]_2\}_2$.

AX_4 Tetrahedral Geometry

In the great majority of its compounds, beryllium adopts the tetrahedral AX_4 geometry in both molecular and extended structures. Beryllium oxide has the wurtzite structure (Figure 4.4) in which both beryllium and oxygen are tetrahedr-

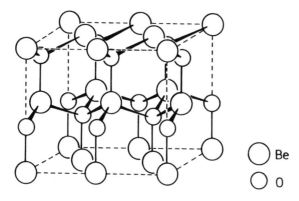

Be

O

Figure 4.4 Wurtzite (ZnS) structure of crystalline beryllium oxide, BeO.

ally coordinated. Beryllium fluoride has the quartz and cristobalite structures of silica, SiO_2, in which beryllium also has the tetrahedral AX_4 geometry (Figure 4.5). Although these compounds are often considered to be ionic, the bonds appear to have considerable covalent character.

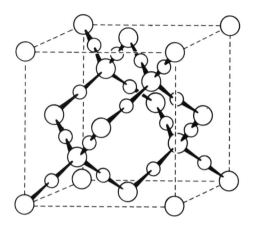

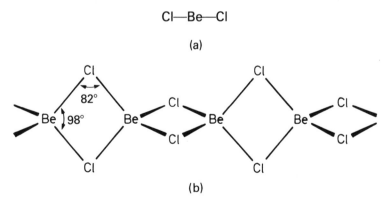

○ Be
○ F

Figure 4.5 Structure of crystalline beryllium fluoride. This structure is the same type as the β-cristobalite form of silica (see Chapter 5).

In the solid state, beryllium dichloride has a chain structure (Figure 4.6) in which there are approximately square four-membered rings in which each beryllium has an approximately tetrahedral AX_4 geometry and each chlorine an angular AX_2E_2 geometry. The bond angle at beryllium is 98.2° and at chlorine only 81.8°. However, it is reasonable to suppose that the bonds in the four-membered ring are bent, as in other small rings, so that the angles between the bonding pairs on each atom are larger than the experimentally observed angles (Figure 4.6), and the four electron pairs in the beryllium valence shell have an arrangement that is much closer to tetrahedral than the experimental bond angles would suggest.

The beryllium chloride chain is broken down by many coordinating ligands to give tetrahedral complexes L_2BeCl_2 such as $(Et_2O)_2BeCl_2$ (Figure 4.7) or ionic

Cl—Be—Cl

(a)

(b)

Figure 4.6 (a) Linear structure of $BeCl_2(g)$. (b) Chain structure of solid beryllium dichloride.

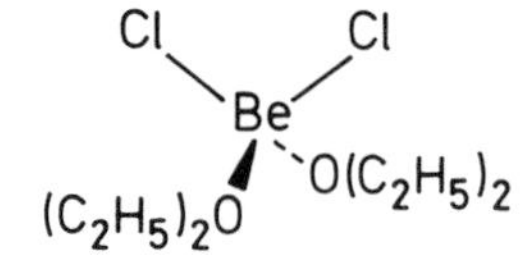

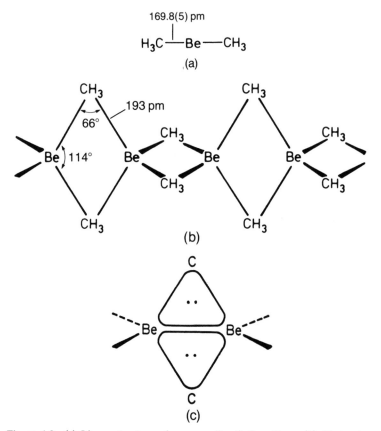

Figure 4.7 Beryllium dichloride dietherate, a tetrahedral AX_4 molecule.

complexes such as $Be(H_2O)_4^{2+}(Cl^-)_2$ and $Be(NH_3)_4^{2+}(Cl^-)_2$ and also by the chloride ion to give the tetrahedral $BeCl_4^{2-}$ ion. The analogous tetrahedral BeF_4^{2-} ion is also well known. Dimethylberyllium has a polymeric structure in the solid state, similar to that of $BeCl_2$, although the bond angles are rather different (Figure 4.8). The differences in the bond angles reflect the bonding in this compound, which is "electron deficient." Each carbon is bonded to two beryllium

Figure 4.8 (a) Linear structure of gaseous dimethylberyllium. (b) Chain structure of solid dimethylberyllium. (c) The bonding may be described in terms of three-center, two-electron bonds.

atoms by one electron pair in a three-center bond. In this case the maximum bonding electron density lies inside, rather than outside, the Be—C direction as in the chloride. There is an approximately tetrahedral arrangement of the four electron pairs around both beryllium and carbon. The Be—C bond is 23 pm longer than in the gas-phase monomer, and this is consistent with the fact that each Be—C bond can be regarded as having a bond order of $\frac{1}{2}$.

Dimethylberyllium dissolves readily in coordinating solvents such as amines. The structure of the 1:2 complex $(CH_3)_2Be.2(quinuclidine)$ in the crystal is shown in Figure 4.9.

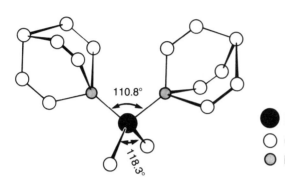

110.8°

118.3°

● Be
○ C
● N

Figure 4.9 Dimethylbis(quinuclidine)beryllium, $(CH_3)_2Be(NC_7H_{13})_2$ (only the nonhydrogen skeleton is shown).

Beryllium forms many complexes with chelating ligands, such as the oxalate ion, $C_2O_4^{2-}$, and 1, 3-diketonates (Figure 4.10) in which there are four tetrahedrally oriented Be—O bonds. Beryllium is unique in forming a series of oxide–carboxylates of general formula $OBe_4(RCO_2)_6$, where R = H, Me, Et, Ph, and so on, of which "basic beryllium acetate" (R = Me) is typical. In these molecules

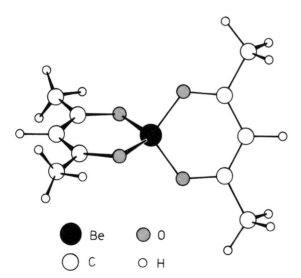

● Be   ● O

○ C   ○ H

Figure 4.10 Bis(acetylacetonato) beryllium, $Be(CH_3COCHCOCH_3)_2$.

there is a central oxygen atom tetrahedrally bonded to four beryllium atoms, each of which also has a tetrahedral AX_4 geometry (Figure 4.11).

In beryllium phthalocyanine (Figure 4.12), beryllium has a square planar coordination that is presumably dictated by the geometry of the planar ligand.

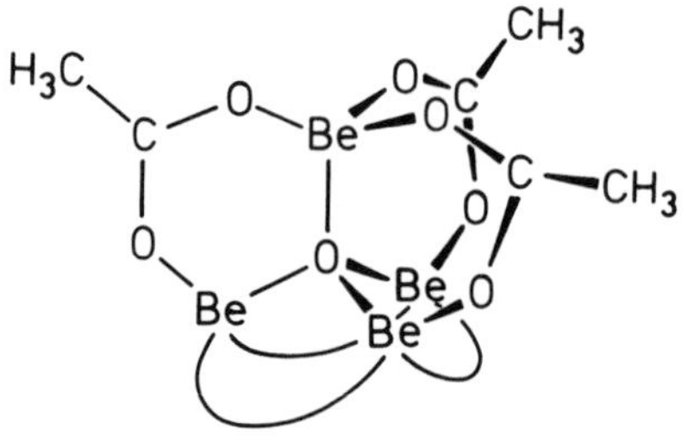

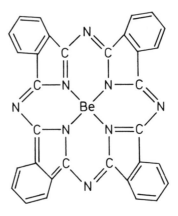

Figure 4.11 Basic beryllium acetate, $OBe_4(CH_3COO)_6$. Three of the acetate groups are represented by curved lines.

Figure 4.12 Beryllium phthalocyanine.

BORON

The only geometries observed for boron are linear AX_2, triangular AX_3, and tetrahedral AX_4. Since boron has only three valence electrons, there are no compounds with a lone pair in the boron valence shell.

AX_2 Linear Geometry

Examples of molecules of boron with this geometry are given in Figure 4.13. Alkali metal borates have stable monomeric molecules in the vapor phase at high temperature with a linear AX_2 geometry at boron. The two BO bonds have lengths of 131 pm and 121 pm, which is consistent with the structure MO—B=O.

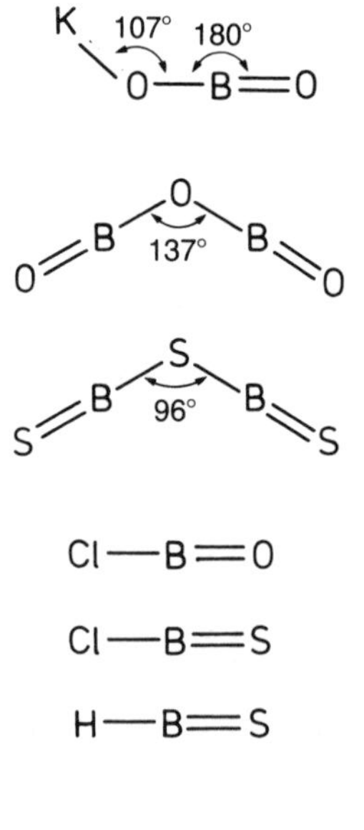

Figure 4.13 Molecules with a linear —B= geometry.

Metaboric acid, HOBO, has a similar structure. The same linear —B= geometry is also found in the oxide and the sulfide, which both have an overall V-shape in the gas phase. Chloro(oxo)boron, chloro(sulfido)boron, and thioborine are also linear (Figure 4.13).

AX₃ Trigonal Planar Geometry

Many simple BX_3 compounds have a planar trigonal geometry. Examples include BF_3, BCl_3, BBr_3, BI_3, $B(CH_3)_3$, $B(CH=CH_2)_3$, $B(NHCH_3)_3$, $B(OCH_3)_3$, and $B(SCH_3)_3$. Because there are three identical ligands in each of these molecules, the bond angles are 120° in each case. Orthoboric acid, $B(OH)_3$, also has a trigonal planar arrangement of the OH groups about the boron atom in the crystal (Figure 4.14). The molecules are held together in sheets by hydrogen bonds.

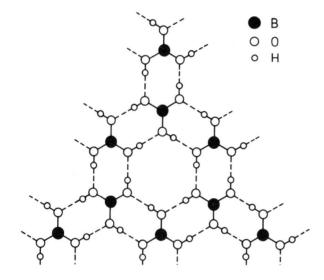

● B
○ O
o H

Figure 4.14 Hydrogen-bonded layer in the structure of crystalline orthoboric acid $B(OH)_3$. Each boron atom has a trigonal planar AX₃ geometry.

When they have been accurately determined, the variations in the bond angles in BX_3 molecules with mixed ligands show the expected trends. For example, the F—B—F angle is smaller than 120° in BF_2X molecules, although the deviations are not large (Table 4.2).

TABLE 4.2 FBF BOND ANGLES AND
B—F BOND LENGTHS IN F₂BX
MOLECULES

F_2BX	FBF (°)	B—F (pm)
X = F	120	131.2(1)
Cl	118.1(5)	131.5(5)
H	118.3(10)	131.1(5)
OH	118.0(10)	132.3(10)
NH₂	117.9(17)	132.5(12)
BF₂	117.2(2)	131.7(2)

The lower halides B_2X_4 also have trigonal planar geometry about the boron atoms (Figure 4.15 and Table 4.3). B_2F_4 has an overall planar geometry with a rather long B—B bond and in this respect it resembles the oxalate ion, $C_2O_4^{2-}$ and dinitrogen tetraoxide, N_2O_4, with which it is isoelectronic. B_2Cl_4 is planar in the

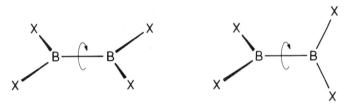

Figure 4.15 Planar and nonplanar forms of the diboron tetrahalides, B_2X_4 (the torsion axis is shown).

TABLE 4.3 DIBORON TETRAHALIDE MOLECULES: XBX BOND ANGLES, B—B BOND LENGTHS, AND BARRIERS TO INTERNAL ROTATION, V_0

Molecule	XBX (°)	B—B (pm)	V_0 (kJ/mol)
B_2F_4	117.2(2)	172.0(4)	2
B_2Cl_4	118.7(3)	170(3)(g), 175(s)	8
B_2Br_4	120.7(3)	168.9(16)	13
B_2I_4	123	169	18

solid state, but in the gas phase it has a staggered conformation with a rather shorter B—B bond. B_2Br_4 has the staggered conformation in both the solid state and the gas phase. It has been estimated that the staggered form of B_2Cl_4 is more stable than the planar form by approximately $8\ kJ\ mol^{-1}$ and that rotation becomes more hindered with decreasing electronegativity of the halogen. Although the VSEPR model can predict the geometry around any one central atom with very few exceptions, it is less reliable for predicting the relative arrangements of the bonds around two adjacent atoms. We would expect that bond-pair bond-pair repulsions would be minimized in the staggered conformation which would therefore be expected to be the most stable, as is the case for $B_2Cl_4(g)$ and B_2Br_4 (g and s), although not for B_2F_4, or for N_2O_4 and $C_2O_4^{2-}$ for the planarity of which no thoroughly convincing explanation has been proposed.

There are several borates in which all the boron atoms have a trigonal planar AX_3 geometry (Figure 4.16), and there are many others in which some

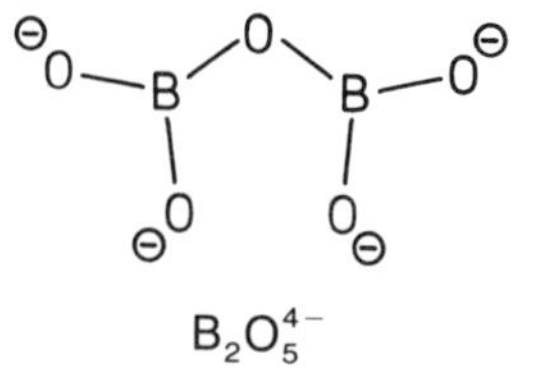

$B_2O_5^{4-}$

Figure 4.16 Borate anion containing boron atoms with a trigonal planar AX_3 geometry.

boron atoms have planar AX_3 geometry while others have a tetrahedral AX_4 geometry (Figure 4.17).

Many boron–nitrogen compounds also have the AX_3 geometry. These include borazine and its many derivatives (Figure 4.18) and boron nitride, which has a planar layer structure closely resembling graphite.

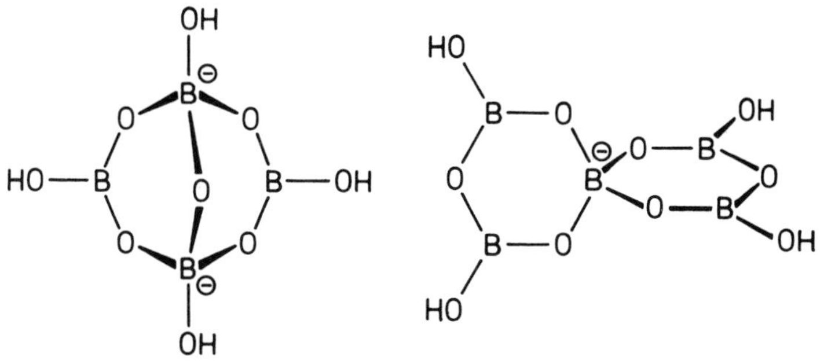

$B_4O_5(OH)_4^{2-}$ $B_5O_6(OH)_4^-$

Figure 4.17 Borate anions containing boron atoms with both a trigonal planar AX_3 and a tetrahedral AX_4 geometry.

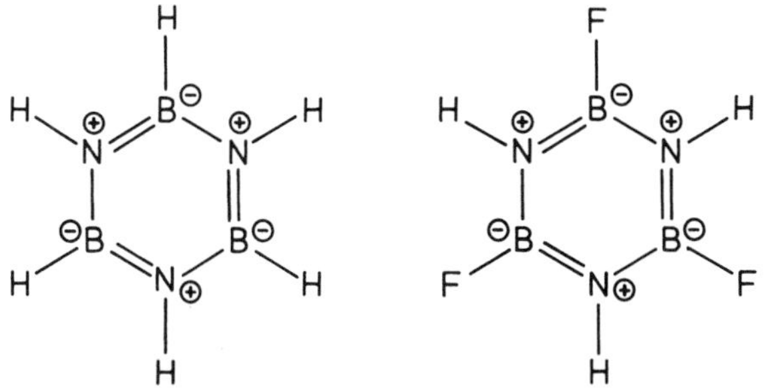

Figure 4.18 Boron atoms in borazine and its derivatives have a trigonal planar AX_3 geometry.

Tetrahedral AX_4 Geometry

There are many borates that have boron atoms with a tetrahedral AX_4 geometry. The simplest of these are the BO_4^{5-} and $B(OH)_4^-$ ions. Other examples are given in Figures 4.19 and 4.17.

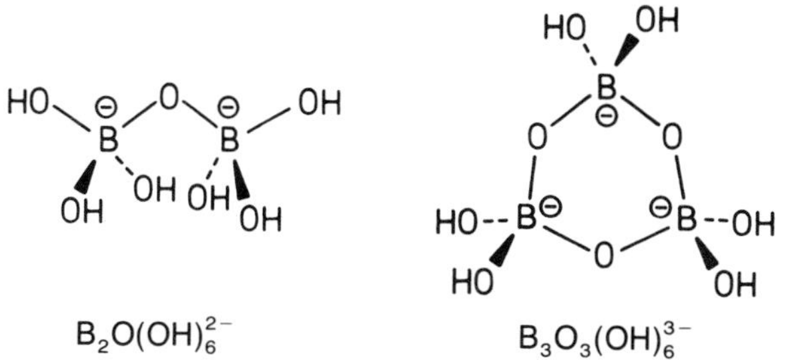

$B_2O(OH)_6^{2-}$ $B_3O_3(OH)_6^{3-}$

Figure 4.19 Borate anions containing boron atoms with a tetrahedral AX_4 geometry.

Borane, BH_3, trimethylborane, $B(CH_3)_3$, and the boron trihalides, BX_3 (X = F, Cl, Br, I), often act as acceptors in donor–acceptor complexes, $X_3B.D$, in which the boron atom acquires a tetrahedral AX_4 geometry. The donor molecules include many in which the donor atom is oxygen, nitrogen, or phosphorus (Figure 4.20). The donor may also be a halide ion or the hydride ion,

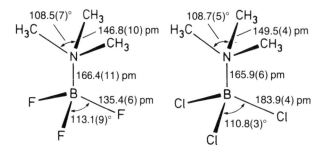

Figure 4.20 $X_3B.D$ donor–acceptor complexes in which the boron atom has a tetrahedral AX_4 geometry.

giving tetrahedral AX_4^- ions such as BH_4^- and BF_4^-. The B—X bonds are markedly longer in the complexes than in the BX_3 molecule. This lengthening may amount to 7 to 8 pm for B—F bonds and 8 to 12 pm for B—Cl bonds. This difference in bond lengths very probably reflects the stronger bond-bond repulsions in AX_4 molecules than in AX_3 molecules.

Boranes and Other Electron-deficient Molecules

Electron-deficient molecules cannot be described in terms of electron-pair bonds. It is necessary to invoke multicenter bonds or to use a molecular orbital description, and so it might appear that the VSEPR model is not applicable to these molecules. However, for those molecules that can be described in terms of three-center bonds the VSEPR model can be used very satisfactorily to rationalize the observed geometry. Thus in diborane and tetramethyldiborane (Figure 4.21) the BHB bridge bonds can be described as three-center bonds in which one electron pair binds three nuclei (Figure 3.16). In these two molecules there is then a tetrahedral arrangement of four electron pairs around each boron, two of which form two-center terminal B—H or B—C bonds and two of which form three-center bridge bonds. Because an electron pair forming a three-center bond is shared between the H atom and two boron atoms, it occupies a smaller domain in the valence shell of each boron atom than the electron pair of a terminal bond. Thus the angle at boron between the two bridge bonds is smaller than that between the terminal bonds. The higher boranes can only be satisfactorily described in terms of molecular orbitals or three-center, four-center, and other multicenter bonds and the VSEPR model is no longer useful. But if the electronic

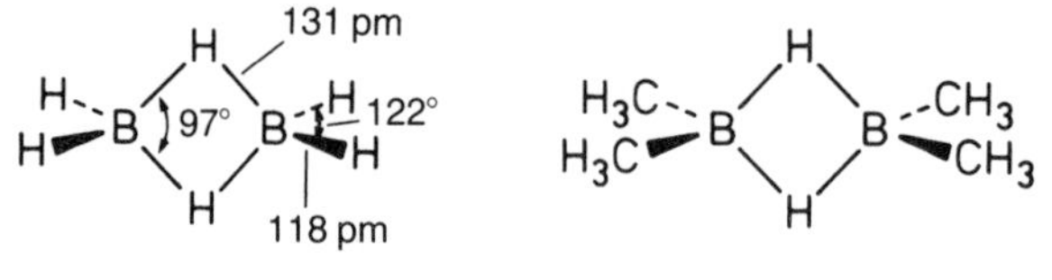

Figure 4.21 Diborane and tetramethyldiborane. The bridge bonds are three-center, two-electron bonds.

structure of a molecule can be described in terms of two- and three-center bonds only, then the VSEPR model can be used to rationalize molecular geometry. For example, the molecule B_4Cl_4 has a tetrahedral cage of boron atoms (Figure 4.22). If it is considered that there is an electron pair in each face of the tetrahedron, thus forming a three-center bond with the three boron atoms at the corners of the face, then there is an approximately tetrahedral arrangement of four electron pairs in the valence shell of each boron atom.

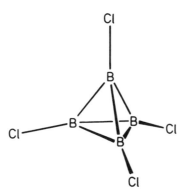

Figure 4.22 The tetrahedral B_4Cl_4 molecule may be considered to have an electron pair in each face of the tetrahedron forming a three-center, two-electron bond, thus giving a tetrahedral arrangement of four electron pairs around each boron.

CARBON

AX₄ Tetrahedral Geometry

Each carbon atom in diamond has a regular tetrahedral geometry, as does the carbon atom in CX_4 molecules in which all the ligands are identical. Bond angles at carbon in a number of simple CX_4 compounds are listed in Table 4.4. All the deviations from the tetrahedral angle that are observed when the ligands are not identical are very small. This is a consequence of the four bonding pairs of electrons being closely packed in the small valence shell of the carbon atom, so changes from the tetrahedral angle are resisted by strong electron-pair repulsions. When the ligands are hydrogen and fluorine, the angle between the two fluorines is always the smallest, which is consistent with the greater electronegativity of fluorine and the resultant smaller size of the C—F bonding electron pair, and this is also the case when the ligands are fluorine and chlorine. However, when the ligands are chlorine and hydrogen, the chlorine bond angle is the largest, despite the greater electronegativity of chlorine. That the ClCCl bond angles are larger than predicted by the VSEPR model can be attributed to the effects of non-bonded ligand–ligand repulsions. With a small central atom, such as carbon, and a relatively large ligand, such as chlorine, repulsions between the ligands are expected to be of some importance. The chlorine–chlorine nonbonded distances are 293 and 290 pm in CH_2Cl_2 and $CHCl_3$, respectively. Thus they are only slightly larger than twice the 1, 3 nonbonded radius of chlorine (288 pm). The BrCBr angle is also larger than the HCH angle in CH_2Br_2, and this angle is larger than HCBr in both CH_3Br and $CHBr_3$ for similar reasons.

TABLE 4.4 BOND ANGLES (°) AT TETRAHEDRAL CARBON ATOMS

Molecule	HCH	FCH	FCF
CH_3F	110.6(2)	108.3	—
CH_2F_2	112.8(3)	108.9	108.5(1)
CHF_3	—	110.3	108.6(3)

Molecule	HCH	ClCH	ClCCl
CH_3Cl	110.4(1)	108.5	—
CH_2Cl_2	111.5(2)	108.3	112.0(2)
$CHCl_3$	—	107.6	111.3(2)

Molecule	HCH	BrCH	BrCBr
CH_3Br	111.2(1)	107.7	—
CH_2Br_2	110.9(8)	108.3	112.9(2)
$CHBr_3$	—	107.2(4)	111.7(4)

Molecule	FCF	ClCF	ClCCl
CF_3Cl	108.6(2)	110.3	—
CF_2Cl_2	106.2(4)	109.5	112.6(5)
$CFCl_3$	—	108	111(1)

AX_3E Trigonal Pyramidal Geometry

Carbanions CX_3^- are expected to have a trigonal pyramidal geometry. However, the only structures that have been determined are those of carbanions with strongly electron withdrawing ligands, such as NO_2 and CN. These anions have a planar structure, for example, $C(NO_2)_3^-$, or a nearly planar structure, as is the case for $C(CN)_3^-$ (Figure 4.23). The planarity or near planarity of these carbanions may be ascribed to extensive delocalization of the carbon lone pair into the

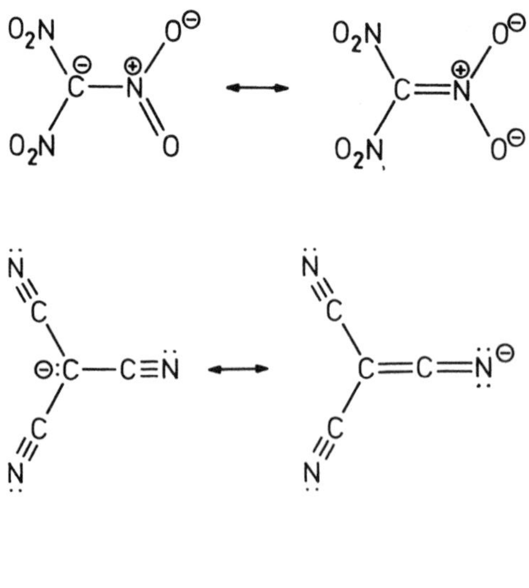

Figure 4.23   The structures of some carbanions R_3C^-. $C(NO_2)_3^-$ is planar and $C(CN)_3^-$ is almost planar because of extensive delocalization of the carbon lone pair as shown by the resonance structures.

valence shells of the ligand atoms. This may be described in terms of the resonance structures shown.

The radicals $\cdot CF_3$ and $\cdot CBr_3$ have a trigonal pyramidal geometry. They can be considered as AX_3e systems, where e represents a single unpaired electron. The observed bond angle is approximately $110°$ in both cases.

AX_3 Trigonal Planar Geometry

Carbocations R_3C^+ are expected to have a trigonal planar AX_3 geometry. There has been no complete determination of the geometry of a carbocation, but vibrational spectroscopic studies of some carbocations in solution are consistent with the expected planar geometry. In the vast majority of molecules in which carbon has this geometry, it is forming one double bond and two single bonds, or three bonds of intermediate bond order, that have a total bond order of four, as in the carbonate ion for example (Figure 4.24). Tables 4.5 and 4.6 give data for a series of $X_2C{=}O$ and $X_2C{=}CH_2$ molecules, respectively. The $={=}C{-}$ angles are

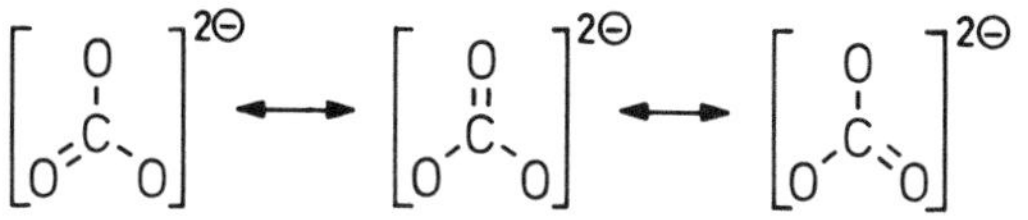

Figure 4.24 Resonance structures for the carbonate ion.

TABLE 4.5 GEOMETRICAL PARAMETERS FOR $X_2C{=}O$ MOLECULES

$X_2C{=}O$	XCX (°)	XCO (°)	CO (pm)
X = H	116.5(7)	121.7	120.8(3)
F	107.7(1)	126.2	117.2(1)
Cl	111.8(1)	124.1	117.6(3)
Br	112.3(4)	123.8	117.8(9)
CH_3	116.0(3)	122.0(3)	121.4(4)

TABLE 4.6 GEOMETRICAL PARAMETERS FOR $X_2C{=}CH_2$ MOLECULES

$X_2C{=}CH_2$	XCX (°)	XCC (°)	C=C (pm)
X = F	110.6	124.7(3)	134.0(6)
Cl	114.1(2)	123.0	133.4(4)
CH_3	115.6(2)	122.2	134.2(3)

consistently larger than the $-C-$ angles. The XCX angle shows the expected dependence on ligand electronegativity so that the $F-C-F$ angle is the smallest in both series. The $C{=}O$ bond length decreases with increasing electronegativity of X. As the space occupied by the $C-X$ bond-pair domain decreases with increasing electronegativity of X, the two electron pairs of the double bond can spread out a little and move close to the core of the carbon atom, thus decreasing the $C{=}O$ bond length. In the $X_2C{=}CH_2$ series the variation in the XCX angles is smaller than in the $X_2C=O$ series and no significant variation can be observed in the $C=C$ bond length.

Each carbon atom in the benzene molecules has a planar AX_3 geometry in which the CCC and CCH angles are all 120°. But in a substituted benzene the CCC angle adjacent to the substituent, called the *ipso* angle, depends on the electronegativity of the substitutent. On substituting a hydrogen by a more electronegative ligand, such as fluorine, the weaker repulsion between the C—F bonding pair and the adjacent CC bond pairs leads to an increase in the ipso angle. Conversely, a less electronegative substituent, like the methyl group, decreases the ipso angle (Figure 4.25 and Table 4.7). The lengths of the two CC bonds forming the CCC angle also change as expected. For angles smaller than 120°, these CC bonds are longer than in benzene, where they have a length of 139.7 pm, and for angles greater than 120°, they are shorter.

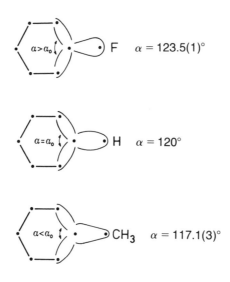

Figure 4.25 Variation of the ipso angle (α) of the benzene ring with the electronegativity of the substituent.

TABLE 4.7 RING GEOMETRY OF MONOSUBSTITUTED BENZENE DERIVATIVES

Parameters	Li[a]	CH$_3$[b]	F[b]
Angles (°):			
α	111.2	118.7(4)	123.4(2)
β	125.2	120.4(4)	118.0(2)
γ	120.0	120.6(5)	120.2(3)
δ	118.6	119.4(6)	120.2(4)
Bond lengths (pm):			
a	142.3	140.8(5)	138.7
b	139.3	139.6(4)	139.9
c	138.7	139.6(4)	140.1

[a]Crystalline $(C_6H_5LiOC_2H_5)_4$.
[b]Gaseous.

Graphite consists of parallel planar layers of carbon atoms that are only weakly bonded together. In each layer each carbon atom is bonded to three others by two single and one double bond (Figure 4.26). The trigonal planar arrangement of these three bonds determines the planarity of each carbon atom

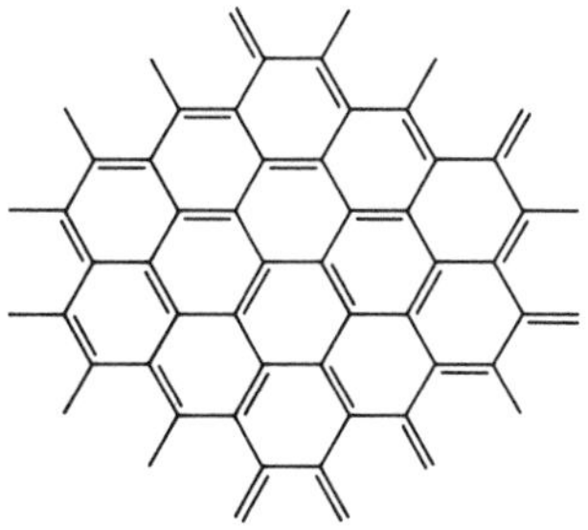

Figure 4.26 One resonance structure for a single layer of graphite. Each carbon atom has a trigonal planar AX_3 geometry.

layer. In fact, all three bonds are equivalent and all the bond angles are equal to 120° as each double bond can occupy each of the three possible positions on each carbon atom; in other words each bond can be considered to have a bond order of $1\frac{1}{3}$. This is equivalent to having one electron from each carbon atom in delocalized orbitals covering the whole of each layer. However, this does not change the geometry, which is determined by the planar arrangement of the three localized bonds at each carbon atom.

AX_2E Angular Geometry

Methylene, CH_2, has an angular geometry with a bond angle of 102.4°. Difluoromethylene, CF_2, has a slightly larger bond angle of 104.8(1)° and CBr_2 has an angle of 114°. Calculations have given a value of 109° for CCl_2. All these CX_2 molecules are unstable and difficult to study. The bond angle is expected to be smaller than 120° in all these molecules and is expected to increase with decreasing electronegativity of the ligand, as is observed for CF_2, CCl_2, and CBr_2, although the angle observed for CH_2 is unexpectedly small. It is often found that the bond angle between two X—H bonds is smaller than would be expected from the electronegativity of hydrogen. Because a hydrogen atom has no other electron pairs in its valence shell, the domain of the bonding pair must also surround the nucleus; so the space that it occupies in the valence shell of an atom to which it is attached is smaller than would be expected from its electronegativity and this may be the reason why some HXH bond angles are smaller than expected.

AX_2 Linear Geometry

When carbon forms two double bonds or a single bond and a triple bond, it is expected to have a linear AX_2 geometry, as is in fact observed. Some examples are given in Figure 4.27.

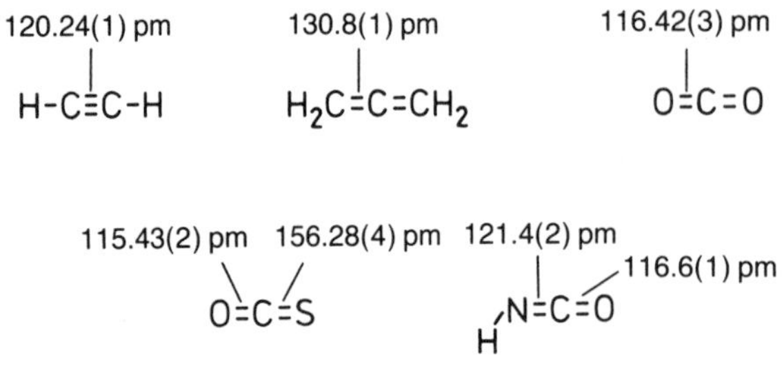

120.24(1) pm   130.8(1) pm   116.42(3) pm

H-C≡C-H $H_2C \!\!=\!\! C \!\!=\!\! CH_2$ O=C=O

115.43(2) pm   156.28(4) pm   121.4(2) pm

                         116.6(1) pm

O=C=S   N=C=O, H

Figure 4.27 Some molecules with a linear AX_2 geometry at carbon.

NITROGEN

AX_4 Tetrahedral Geometry

The expected tetrahedral arrangement for four single bonds around a positively charged nitrogen atom has been established for the ammonium ion, some tetraalkyl ammonium ions, and the NF_4^+ ion.

    In AX_4 molecules in which there are nonequivalent ligands the expected deviations from the ideal tetrahedral angle are observed. Figure 4.28 gives the bond angles in ONF_3 and $ON(CH_3)_3$. We see that, although the bond angles are nearly ideal tetrahedral in the methyl derivative, the FNF angles are much smaller and, accordingly, the ONF angle is much larger than 109.5° in ONF_3. The smaller

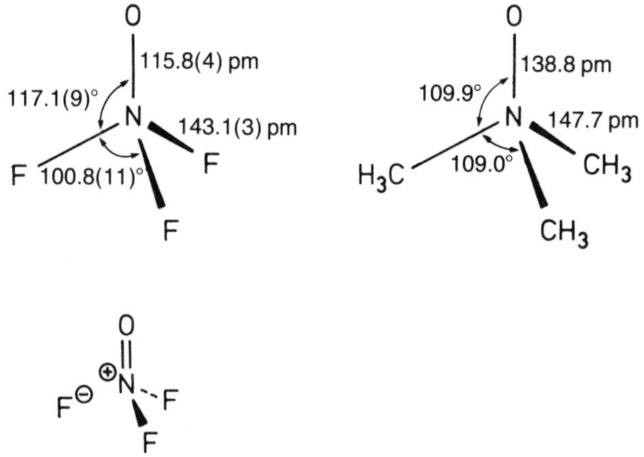

O   O

115.8(4) pm   138.8 pm

117.1(9)°   N   109.9°   N

        143.1(3) pm   147.7 pm

F 100.8(11)°   F H_3C 109.0° CH_3

       F CH_3

O

F⁻   ⊕N —F

     F

Figure 4.28 Tetrahedral AX_4 structures of ONF_3 and $ON(CH_3)_3$.

FNF angle is consistent with the increase in the ligand electronegativity on going from methyl to fluorine. Removal of electron density from the nitrogen atom by the electronegative fluorines allows a greater delocalization of the formally lone-pair electrons on oxygen into the valence shell of the nitrogen. This delocalization may be described by resonance structures such as that shown in Figure 4.28. The NO bond has more double-bond character in ONF_3 than in $ON(CH_3)_3$ and is shorter than the NO bond in $ON(CH_3)_3$, and because its domain is larger it occupies more space in the valence shell and thus decreases the

FNF bond angles. The NF bonds are longer, and the FNF angles smaller, in ONF_3 than in NF_3 [136.5(2) pm, 102.5°], which is consistent with the NO bond domain being larger than the lone-pair domain in NF_3.

AX₃E Trigonal Pyramidal Geometry

The bond angle of ammonia, NH_3 (107.2°), is somewhat smaller than the ideal tetrahedral angle as a consequence of the presence of a lone pair. The bond angle of trifluoroamine, NF_3, is considerably smaller, 102.3(3)°, which is consistent with the greater electronegativity of fluorine. On the other hand, the bond angle in NCl_3 is 107.1(5)°, which is the same, within experimental error, as the bond angle of ammonia, although it is expected to be intermediate between those of NF_3 and NH_3. The nonbonded Cl . . . Cl distance in NCl_3 is 283.0(2) pm, which is 5 pm shorter than twice the chlorine 1, 3 nonbonded radius (288 pm). Thus the bond angle in NCl_3 appears to be limited by repulsions between the Cl atoms, which is confirmed by the bond angles in NH_2Cl, which are HNH, 107(2)°, and HNCl, 103.7(2)°. Thus the HNCl angle is smaller than both HNH in NH_3 and ClNCl in NCl_3, indicating that the bond angle in NCl_3 would be smaller if this was not prevented by nonbonded repulsions.

The bond angles in $N(CH_3)_3$ and $N(CF_3)_3$, 110.9(6)° and 117.9(4)°, respectively (Figure 4.29), are both larger than the tetrahedral angle, again presumably because of the bulkiness of the ligands. Although the CF_3 group is more electronegative than the CH_3 group, the bond angle in $N(CF_3)_3$ is even larger

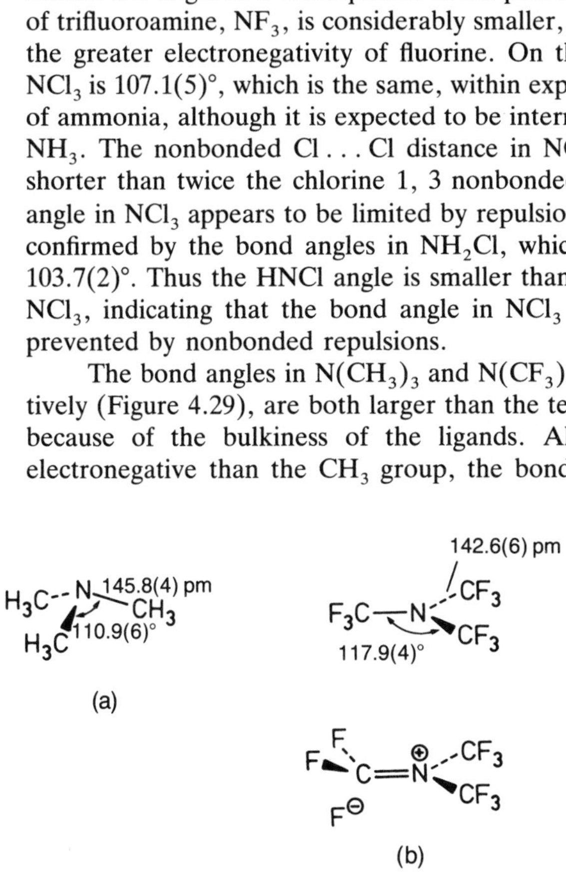

(a)

(b)

Figure 4.29 Pyramidal AX₃E molecules: (a) Trimethylamine. (b) Tris(trifluoromethyl)amine is almost planar.

than in $N(CH_3)_3$. The CN bonds in $N(CF_3)_3$, which have a length of 142.6(6) pm, are considerably shorter than in $N(CH_3)_3$, 145.8(4) pm. This suggests that there is some delocalization of the nitrogen lone pair onto the carbon atoms, which can be represented by resonance structures such as that shown in Figure 4.29. This lone-pair delocalization is presumably at least partly responsible for the large bond angle in $N(CF_3)_3$.

Similar trends are observed in tetrafluorohydrazine, $F_2N—NF_2$, which occurs in two torsional forms, and tetrakis(trifluoromethyl)hydrazine, $(CF_3)_2N—N(CF_3)_2$ (Figure 4.30). In both forms of N_2F_4 the nitrogen atoms have

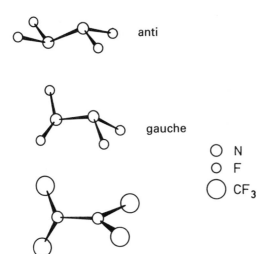

anti

gauche

O N
O F
O CF₃

Figure 4.30 Two conformations "anti" and "gauche" are in equilibrium in the vapor of N_2F_4. In both forms the nitrogens have a pyramidal AX_3E geometry. $N_2(CF_3)_4$ is known only in the gauche conformation.

a pyramidal AX_3E geometry with an average angle of 102°. In $(CF_3)_2N\!-\!N(CF_3)_2$ the N—N bond is much shorter, 140(2) pm, than in N_2F_4, 149.2(7) pm. This short N—N bond is consistent with a delocalization of the lone-pair electrons on nitrogen, as in $N(CF_3)_3$ (Figure 4.29), which therefore reduces the repulsion between these two lone pairs.

Tris(trifluoromethylthio)amine, $N(SCF_3)_3$, has a nearly planar geometry at nitrogen, and trisilylamine, $N(SiH_3)_3$, is planar within experimental error (Figure 4.31). In trisilylamine the N—Si bond length of 173.42 pm is much smaller than

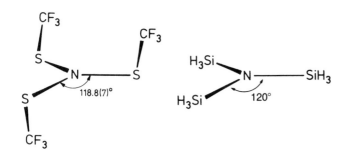

F₃CS⟍ ⊕ ⊖
     N=SCF₃
F₃CS⟋

H₃Si⟍ ⊕ ⊖
     N=SiH₃
H₃Si⟋

Figure 4.31 Tris(trifluoromethylthio)-amine and trisilylamine have nearly planar and planar geometries, respectively.

the value of 181 pm predicted for a single bond. Thus it again seems probable that there is considerable delocalization of the lone pair on nitrogen, as indicated in Figure 4.31. However, it is also important to note that in all NX_3 molecules, except NH_3 and NF_3, the nonbonded X . . . X distances are approximately twice the 1, 3 nonbonded radius of X so that ligand–ligand repulsions are probably also important. In this connection it is noteworthy that in $HN(SiH_3)_2$ the NSiN angle increases to 127.7(3), which gives an Si . . . Si distance of 310 pm, which is exactly twice the 1, 3 nonbonded radius for silicon.

AX_2E_2 Angular Geometry

NX_2^- ions are expected to have the angular AX_2E_2 geometry. Although no structures of such ions have been determined experimentally, the bond angles of NF_2^- and NH_2^- have been found to be 99.4° and 96.7° from ab initio calculations. Again we see, as for CF_2 and CH_2, that the angle between hydrogens is smaller than the angle between fluorines, although the opposite would have been expected from their relative electronegativities. The radicals $\cdot NH_2$ and $\cdot NF_2$ can be considered to be X_2NEe molecules with one lone pair and one single nonbonding electron on nitrogen. The bond angles in $\cdot NH_2$ and $\cdot NF_2$ are 103.4(5)° and 103.2(1)°, respectively. Because of the smaller domain occupied by the single unpaired electron, they have larger bond angles than the corresponding NX_2^- ions.

AX_3 Trigonal Planar Geometry

When nitrogen forms only three bonds and has no lone pair, the three bonds always adopt the expected trigonal planar arrangement. Trisilylamine and other molecules in which it is considered that the nitrogen lone pair is extensively delocalized, and which may therefore be described by resonance structures in which the nitrogen forms two single and one double bond, may also be considered to belong to this class of molecule (Figure 4.31).

The nitrogen trioxide molecule and the nitrate ion both have trigonal planar structures and both may be described by three equivalent resonance structures (Figure 4.32). They have bond angles of 120° and bond orders of $1\frac{1}{3}$. In nitric acid

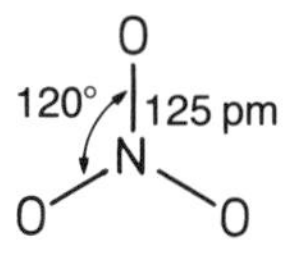

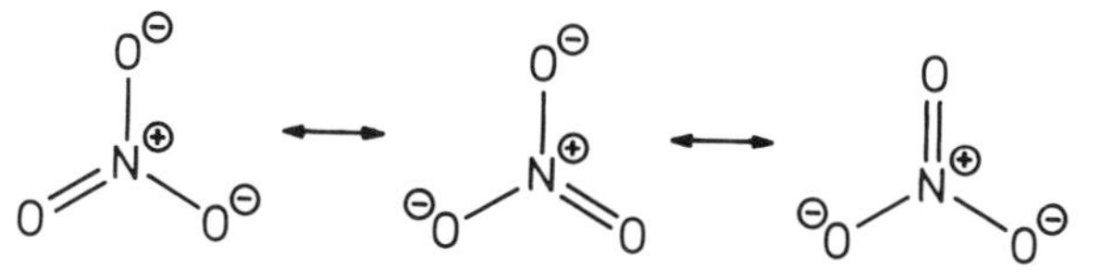

Figure 4.32 Geometry and resonance structures for the nitrate ion.

(Figure 4.33) there are two short NO bonds, which have lengths of 120.3(3) and 121.0(3) pm and bond orders of 1.5, and a single bond of length 140.6(3) pm. As expected, the angle of 130.0° between the two shorter bonds is greater than the ideal angle of 120°, while the other two angles of 116.1(3)° and 113.9(3)° are smaller than the ideal angle.

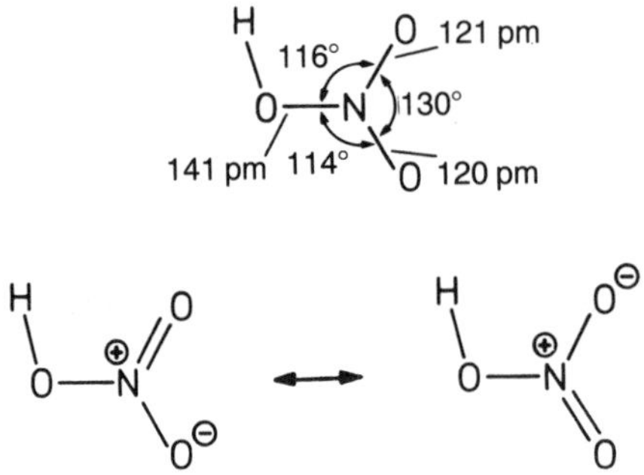

Figure 4.33 Geometry and resonance structures for the nitric acid molecule.

The AX$_3$ planar geometry is also found in dinitrogen trioxide, N$_2$O$_3$, dinitrogen tetraoxide, N$_2$O$_4$, and dinitrogen pentaoxide, N$_2$O$_5$ (Figure 4.34). The resonance structures for these molecules indicate that, with the exception of the NO bridge bonds in N$_2$O$_5$, the NO bonds have a bond order of 1.5, and therefore the angles between these bonds are in every case greater than the ideal angle of

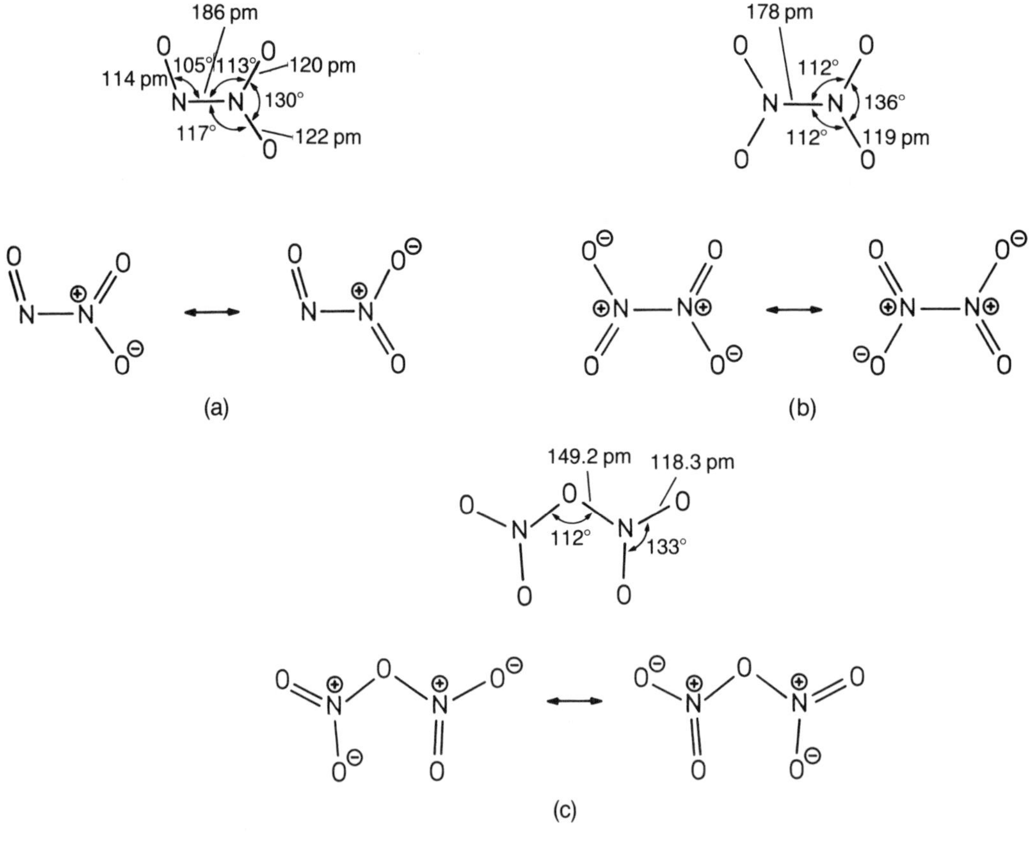

(a)

(b)

(c)

Figure 4.34 Geometry and resonance structures for (a) N$_2$O$_3$, (b) N$_2$O$_4$, and (c) N$_2$O$_5$.

120°. Nitryl fluoride and nitryl chloride have similar planar structures (Figure 4.35), and consistent with the greater electronegativity of fluorine the FNO angle (112°) is smaller than the ClNO angle (114.5°).

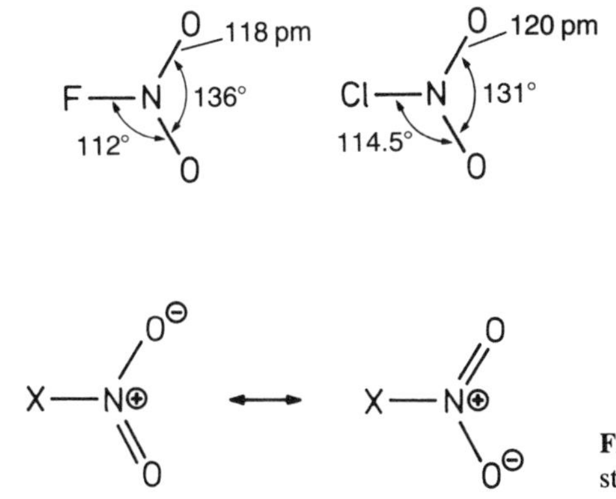

O 118 pm

F—N 136°

112° O

O 120 pm

Cl—N 131°

114.5° O

X—N⊕ ⟷ X—N⊕

Figure 4.35 Geometry and resonance structures for FNO_2 and $ClNO_2$.

AX₂E Angular Geometry

In this class of molecule there is a single bond domain, a double-bond domain, and a lone pair in the valence shell of nitrogen. The nitrosyl halides have a bond angle of less than 120°, which decreases with increasing electronegativity of the halogen: BrNO 114.5(10)°, ClNO 113.3(1)°, and FNO 110.1(5)°. In nitrous acid, HONO, the ONO bond angle is 114° in the syn form and 111° in the antiform (Figure 4.36).

Dinitrogen difluoride, N_2F_2, and azomethane, $N_2(CH_3)_2$, are known in both *cis* and *trans* forms (Figure 4.37). The bond angle is 114° in *cis* N_2F_2 and 106° in *trans* N_2F_2. The hyponitrite ion $N_2O_2^{2-}$ is isoelectronic with N_2F_2 and has been found by vibrational spectroscopy to have the *trans* structure.

Nitrogen dioxide, NO_2, can be considered to be an example of an AX_2e molecule if the two resonance structures given in Figure 4.38 are considered to be the most important. The bond angle of 134.3° is larger than 120° because of the

F—N≋O   Cl—N≋O   Br—N≋O

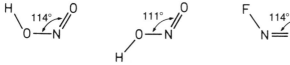

Figure 4.36 Angular X₂NE molecules. The ONO bond angle is different in the syn and anti conformers of nitrous acid.

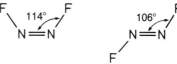

Figure 4.37 *Cis* and *trans* forms of dinitrogendifluoride.

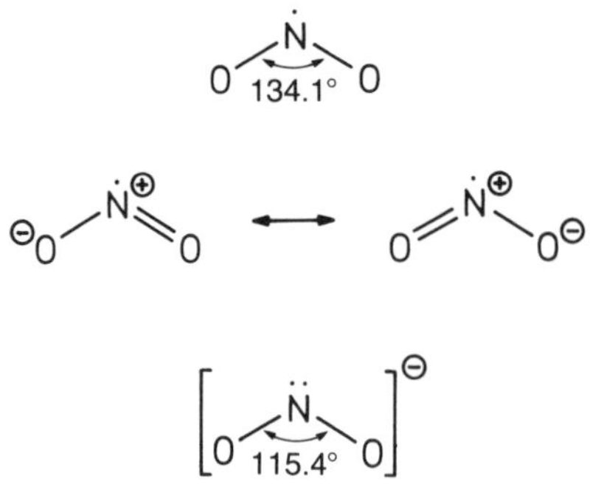

Figure 4.38 Nitrogen dioxide, an angular AX_2e molecule, and the nitrite ion, an AX_2E molecule.

small size of the single-electron domain on nitrogen. In contrast, the bond angle in the nitrite ion, NO_2^-, in which there is a lone pair on nitrogen is only 115.4° (Figure 4.38).

Isocyanates, X—N=C=O, and isothiocyanates, X—N=C=S, have a large range of bond angles at nitrogen. Table 4.8 lists a few examples. There have been many investigations of the structures of silyl isocyanate, H_3SiNCO, and silyl

TABLE 4.8 NITROGEN BOND ANGLES IN ISOCYANATES AND ISOTHIOCYANATES

Isocyanates	XNC (°)	Isothiocyanates	XNC (°)
ClNCO	118.8(5)	—	—
HNCO	128.1(5)	HNCS	135.0(2)
ClO_2SNCO	122.4	—	—
F_2PNCO	130.6(8)	F_2PNCS	140.5(7)
F_3SiNCO	160.7(12)	—	—
Cl_3SiNCO	138.0(4)	—	—
H_3SiNCO	150–180	H_3SiNCS	150–180

isothiocyanate, H_3SiNCS, because different techniques have yielded widely different values for the SiNC angle in the range from 150° to 180°. These discrepancies arise because these molecules have large-amplitude, low-frequency deformation vibrations, and so their average geometry may be markedly different from their equilibrium geometry, as explained in Chapter 2. That the bond angles are larger than 120° may be attributed to delocalization of the nitrogen lone pair as expressed by resonance structures such as H—N≡C—S̄ and H_3S̄i=N=C=O.

AX_2 Linear Geometry

Nitrous oxide, N_2O, the nitronium ion, NO_2^+, the azide ion N_3^-, the cyanamide ion, NCN^{2-}, and the cyanate ion, NCO^-, are isoelectronic and they all have a linear AX_2 geometry (Figure 4.39). The analogous thiocyanate ion NCS^- is also linear.

113 pm 118 pm   115 pm   118 pm

$\overset{\ominus}{N}=\overset{\oplus}{N}=O$ $O=\overset{\oplus}{N}=O$ $\overset{\ominus}{N}=\overset{\oplus}{N}=\overset{\ominus}{N}$

$\overset{\ominus}{N}=C=\overset{\ominus}{N}$ $\overset{\ominus}{N}=C=O$ $\overset{\ominus}{N}=C=S$ **Figure 4.39** Linear AX$_2$ molecules of nitrogen.

Hydrogen azide, chlorine azide (Figure 4.40), and methyl azide have a linear, or nearly linear, AX$_2$ nitrogen and an AX$_2$E nitrogen with an angular geometry.

Cl   H
$\diagdown$ 108.7(7)° 113 pm $\diagdown$ 109(4)° 113 pm
  N$=$N$=$N   N$=$N$=$N
125 pm$\diagup$ 171.9(7)°   124 pm$\diagup$ 171(5)°   **Figure 4.40** Structures of two azides.

OXYGEN

AX$_4$ Tetrahedral Geometry

This is a rather unusual stereochemistry for oxygen in discrete molecules, but it is found in the oxotetracarboxylates of beryllium such as oxohexaacetatotetraberyllium, Be$_4$O(CH$_3$COO)$_6$, which we mentioned earlier (Figure 4.11). As the oxygen atom in these molecules has a formal double-positive charge, the bonds presumably have considerable ionic character, so the actual charge on the oxygen is considerably smaller.

Many metal oxides have infinite lattice structures with the same tetrahedral arrangement of four bonds around oxygen. Although the bonds have considerable ionic character, there is nevertheless a tetrahedral arrangement of four electron pairs around oxygen. For example, BeO and ZnO have the structure shown in Figure 4.4, and PtO and PdO have the structure shown in Figure 4.41. The structure of PtO and PdO differs considerably from the structure of BeO and

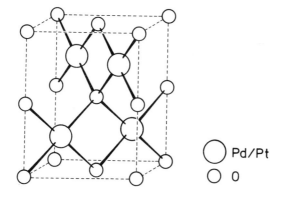

◯ Pd/Pt

◯ O

Figure 4.41 Structures of crystalline platinum and palladium oxides.

ZnO because the metal atoms form four coplanar bonds, rather than four tetrahedral bonds (see Chapter 6).

AX₃E Trigonal Pyramidal Geometry

The hydronium ion, H_3O^+, has a pyramidal AX_3E geometry, but the bond angle varies rather widely with the extent to which the hydronium ion is hydrogen bonded to other ions in the solid. For example, it is 117° in the chloride, 112° in the nitrate, and 101°, 106°, and 126° in the hydrogen sulfate.

Ethers such as dimethyl ether $(CH_3)_2O$ combine with many acceptor molecules such as BF_3 to form complexes such as $(CH_3)_2O.BF_3$. In this particular complex the oxygen has the expected AX_3E geometry (Figure 4.42), although the parameters are not known very accurately. In contrast, in the dimethyl ether–trimethyl aluminum complex (Figure 4.42) the oxygen has trigonal planar geometry. Oxygen has a similar planar geometry in the ion $(HgCl)_3O^+$ and in the

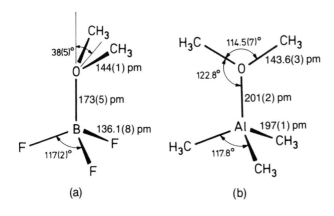

(a)   (b)

Figure 4.42 (a) Dimethylether-boron trifluoride with a pyramidal AX_3E geometry at oxygen. (b) Dimethyl ether-trimethyl aluminum with a trigonal planar geometry at oxygen.

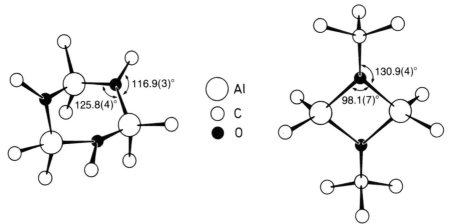

Figure 4.43 Trimeric dimethylaluminummethoxide $[(CH_3)_2AlOCH_3]_3$ and dimeric dimethylaluminum-t-butoxide $[(CH_3)_2AlOC(CH_3)_3]_2$. Both have a trigonal planar AX_3 geometry at oxygen. Only the nonhydrogen skeleton is shown.

molecules $[(CH_3)_2AlOCH_3]_3$ and $[(CH_3)_2AlOC(CH_3)_3]_2$ (Figure 4.43). Delocalization of the oxygen lone pair seems to be the most reasonable explanation for the planarity of the bonds to oxygen in these latter molecules. Accordingly, the geometry at oxygen in these molecules may be considered to be AX_3 trigonal planar.

AX_2E_2 Angular Geometry

The molecules F_2O, FOH, and H_2O have the expected bond angles of less than 109.5°, the F_2O and FOH angles being smaller than the H_2O angle because of the greater electronegativity of fluorine (Figure 4.44). The experimental bond angles

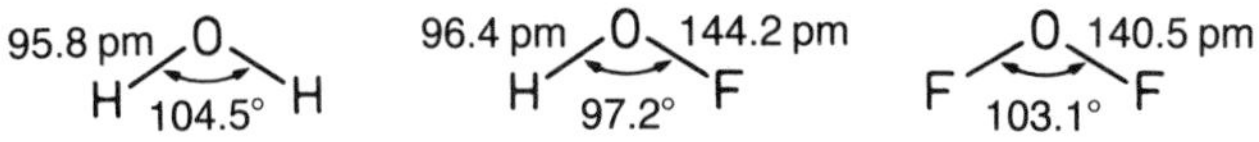

Figure 4.44   Angular OX_2E_2 molecules.

are given in Table 4.9 along with those of some other molecules. It might seem reasonable to suppose that the bond angle in FOH would be intermediate between those of F_2O and H_2O rather than smaller than both. However, the OF bond in FOH will be more polar than in F_2O, as is indicated by the increase in its

TABLE 4.9   OXYGEN BOND ANGLES IN OX_2 MOLECULES

Molecule	Angle (°)
H_2O	104.5(1)
F_2O	103.1(1)
FOH	97.2(6)
Cl_2O	110.9(1)
HOCl	102.5(4)
HO(OH)	95(2)
FO(OF)	109.5(5)
$(CH_3)_2O$	111.8(2)

length from 140.5 to 144.2 pm, so the bonding pair will occupy less space in the valence shell of oxygen. The OH bond will also be more polar than in H_2O, but in the opposite sense, so the bonding pair will occupy more space in the valence shell of oxygen. Thus it is not possible to make a certain prediction about the size of the bond angle in FOH. It is also possible that the attraction between the positively charged hydrogen and a lone pair on the fluorine is sufficiently strong to decrease the bond angle.

The bond angle in Cl_2O is larger than in H_2O rather than smaller, as expected on the basis of electronegativity. This large bond angle may be attributed to nonbonded repulsions, as the Cl...Cl nonbonded distance of 279 pm is smaller than twice the 1, 3 nonbonded radius of chlorine (288 pm). But the bond angle in ClOH is smaller than that in H_2O, just as the bond angle in

FOH is smaller than that in H_2O, because nonbonded repulsions are not important in this case.

In hydrogen peroxide (Figure 4.45) the bond angle of 95° is considerably smaller than in the water molecule, which is consistent with the greater electronegativity of the OH group compared to H. At first sight it seems surprising that the bond angle is smaller than in F_2O, but it is consistent with the O—O bond being unusually long and weak and therefore exerting a weak repulsion on the O—H bond. In contrast, in F_2O_2, where the O—O bond is exceptionally short and strong, the bond angle is correspondingly large. The length and strength of the O—O bond is consistent with resonance structures such as that in Figure 4.45.

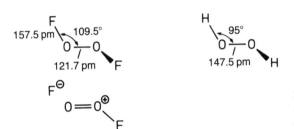

Figure 4.45   Hydrogen peroxide and dioxygendifluoride have an angular AX_2E_2 geometry at oxygen.

In the examples that we have considered so far the bond angle at an AX_2E_2 oxygen is smaller than or close to the ideal tetrahedral angle. However, in many molecules the bond angle at oxygen is considerably larger than the tetrahedral angle, as shown in Table 4.10. These large bond angles are presumably due to

TABLE 4.10 OX_2E_2 MOLECULES IN WHICH THE BOND ANGLE AT OXYGEN IS GREATER THAN 109.5°

$B(OCH_3)_3$		$[(CH_3)_2B]_2O$
BOC	121.4(5)°	BOB 144(3)°
$(CH_3)_3SiOCH_3$		$[(CH_3)_3Si]_2O$
SiOC	122.5(6)°	SiOSi 148(3)°
CH_3OPF_2		$(F_2P)_2O$
POC	123.7(7)°	POP 135.2(9)°
CH_3OSO_2F		$(FO_2S)_2O$
SOC	116.5(7)°	SOS 123.6(12)°

delocalization of the lone-pair electrons on oxygen into the valence shell of one or both of the attached atoms X, where X is boron or an element from the third or subsequent periods. On replacement of the methyl group in CH_3OX by another X, the bond angle increases because of increased lone-pair delocalization.

The SiOSi angles in some disiloxanes are listed in Table 4.11, together with some SiO bond lengths. The angles are generally considerably larger than the tetrahedral angle, and the SiO bonds are shorter than calculated from single bond

TABLE 4.11 SiOSi BOND ANGLES AND Si—O BOND
LENGTHS IN DISILOXANE AND DERIVATIVES

Molecule	SiOSi (°)	SiO (pm)
$H_3SiOSiH_3$	144.1(8)	163.4(2)
$(CH_3)_3SiOSi(CH_3)_3$	148(3)	163.1(3)
$Cl_3SiOSiCl_3$	146(4)	159.2(10)
$F_3SiOSiF_3$	156(2)	158(2)

radii (183 pm), presumably because of delocalization of the oxygen lone pairs.
Thus it is interesting to note that when oxygen is replaced by sulfur the bond
angles are smaller than the tetrahedral angle (Table 4.12). Because sulfur does
not have a filled valence shell, it has little tendency to delocalize its lone pairs.

TABLE 4.12 X—O—X, X—S—X, AND X—Se—X BOND ANGLES,
X...X NONBONDED DISTANCES AND $2r_{1,3}(X)$ VALUES IN A
SERIES OF GASEOUS COMPOUNDS

Molecule	X—O—X (°)	X...X (pm)	$2r_{1,3}(X)$ (pm)
OCl_2	110.9(1)	279	288
$O(CH_3)_2$	111.8(2)	234	250
$O(SiH_3)_2$	144.1(8)	311	310
$O(GeH_3)_2$	126.5(3)	315	316

Molecule	X—S—X (°)	X...X (pm)	$2r_{1,3}(X)$ (pm)
SCl_2	102.7(2)	315	288
$S(CH_3)_2$	99.1(1)	275	250
$S(SiH_3)_2$	98.4(1)	321	310
$S(GeH_3)_2$	98.9(1)	336	316

Molecule	X—Se—X (°)	X...X (pm)	$2r_{1,3}(X)$ (pm)
$Se(CH_3)_2$	96.3(1)	290	250
$Se(SiH_3)_2$	96.6(7)	340	310
$Se(GeH_3)_2$	94.6(5)	345	316

Moreover, ligand–ligand nonbonded repulsions are much less important in these
sulfur compounds. The X...X distances in some X—O—X, X—S—X, and
X—Se—X molecules are given in Table 4.12. We see that the X...X distances in
X—O—X molecules are equal to or smaller than $2r_{1,3}$, whereas the X...X
distances in X—S—X and X—Se—X are invariably larger and very often much
larger than $2r_{1,3}$. The bond angle of $(GeH_3)_2O$ is smaller than in disiloxane. Both
the experimental Si...Si and Ge...Ge distances are equal to twice the 1, 3
nonbonded radius; but since the germanium atom is larger, the bond angle is
smaller. Thus in these molecules the lone pairs decrease the bond angle until the
ligand–ligand repulsions limit any further decrease.

AX₂E Angular Geometry

Ozone has an angular AX_2E geometry with a bond length of 127.2(1) pm and a
bond angle of 117.8(1)°, which is slightly smaller than the ideal angle of 120°

(Figure 4.46). The short bond length is consistent with the bond order of 1.5 indicated by the two resonance structures.

Al_2O and related molecules (Figure 4.47) have the expected angular AX_2E_2 geometry, but the short metal–oxygen bonds and the large bond angles of around 140° indicate that the oxygen lone pairs are probably delocalized. The Ga—O bond, for example, has a length of 182.4(3) pm compared to the sum of the covalent radii, which is 188 pm, and a bond angle of 143(1)°. However, the nonbonded metal–metal distances are approximately equal to, or even shorter than, twice the intramolecular 1, 3 nonbonded radii. For example, the Ga . . . Ga distance is 345 pm, which is essentially the same as $2r_{1,3} = 344$ pm. This suggests that ligand–ligand repulsions cannot be ignored, and they are probably at least partly responsible for the large bond angles.

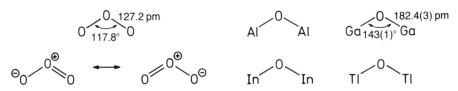

Figure 4.46   Ozone is an angular AX_2E molecule.

Figure 4.47   Some angular M_2O molecules.

FLUORINE

Molecules in which a fluorine atom forms more than one bond are rather rare. In addition to the angular H_2F^+ ion (AX_2E_2), which has a bond angle of 114(2)°, there are two other important types of molecules in which a fluorine atom forms two bonds: (1) the hydrogen-bonded dimer and other polymers of HF that are discussed in the following section, and (2) fluorides that are polymerized by the formation of fluorine bridges in which a fluorine is shared by two monomers. These molecules are discussed in Chapter 5.

HYDROGEN-BONDED COMPLEXES

Hydrogen fluoride and to a lesser extent HCl, HBr, and HCN form hydrogen-bonded complexes with many donor molecules having an unshared pair of electrons, particularly when the lone pair is in the valence shell of a second-period element. The structures of many of these complexes in the gas phase have been determined by microwave spectroscopy. The H—F . . . H—F dimer shown in Figure 4.48(a) is typical. It does not have a structure such as (b) or (c), which might have been expected if the interaction between the two HF molecules was simply the electrostatic interaction between two dipoles. The observed angular structure is most easily understood as resulting from the interaction of the positively charged hydrogen of one HF molecule with one of the unshared pairs of electrons on the fluorine of the other HF molecule. The observed angle of 110° is consistent with an approximately tetrahedral arrangement of the H—F bonding pair and the three lone pairs in the valence shell of the fluorine atom. We will see that in many other cases also a hydrogen bond forms in the direction of a lone pair, assuming that the lone pair and bond pairs in a valence shell have the arrangement postulated by the VSEPR model.

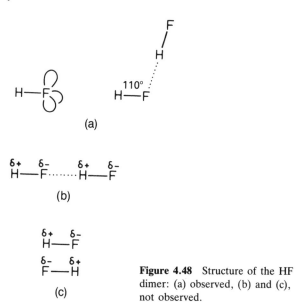

(a)

(b)

(c)

Figure 4.48 Structure of the HF dimer: (a) observed, (b) and (c), not observed.

YAE . . . HX Complexes

In YAE molecules in which the donor atom A has only one attached ligand Y, the lone pair and the ligand Y have a linear arrangement. These molecules therefore form linear hydrogen-bonded complexes. Some examples are shown in Figure 4.49. Other examples include HCN . . . HCl, HCN . . . HBr, HCN . . . HCN, OC . . . HCl, OC . . . HBr, OC . . . HCN, N_2 . . . HCl, N_2 . . . HCN, HC≡C—C≡N . . . HF, and NCCN . . . HF.

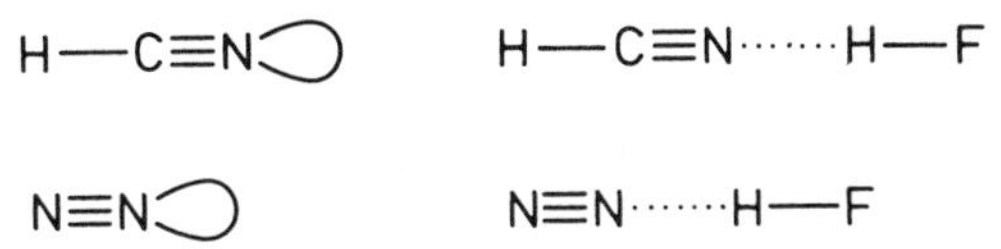

H—C≡N◯   H—C≡N······H—F

N≡N◯   N≡N······H—F

Figure 4.49 Linear YAE . . . HF hydrogen-bonded complexes.

YAE$_2$ . . . HX Complexes

In YAE$_2$ molecules the bond domain and the two lone pairs have a planar triangular arrangement. Thus we expect a hydrogen bond to form at approximately 120° to the bond. In the H_2CO . . . HF complex the observed angle is 110° and in SO_2 . . . HF it is 145° (Figure 4.50).

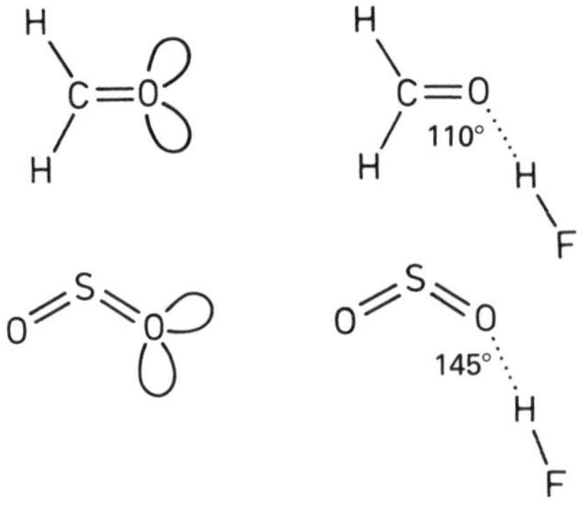

Figure 4.50 Angular YAE$_2$...HF hydrogen-bonded complexes.

YAE$_3$ ... HX Complexes

The HF dimer is an example of this type of complex and, as we have seen, the observed angle is in good agreement with the predicted approximately tetrahedral angle based on the tetrahedral arrangement of three lone pairs and one bond pair (Figure 4.48). The HF ... HCl complex also has the same angular geometry.

Y$_2$AE$_2$ ... HX Complexes

A typical example of this type is the H$_2$O ... HF complex as shown in Figure 4.51. The angle between the plane of the water molecule and the hydrogen bond is 134°, which implies an angle of 115° between the hydrogen bond and one of the

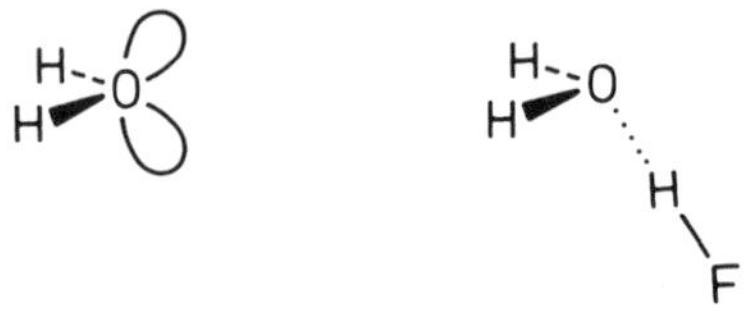

Figure 4.51 Angular Y$_2$AE$_2$ ... HF hydrogen-bonded complexes.

OH bonds. For a tetrahedral arrangement of two lone pairs and two bonding pairs, this angle is expected to be the ideal tetrahedral angle of 109.5°. In the methanol–HF complex the H—O ... H angle is 106.4°. In the H$_2$S ... HF complex the H—S ... H angle is only 91°, but this is consistent with the bond angle in H$_2$S, which is only 92° and which implies that the two lone pairs have an almost collinear arrangement perpendicular to the plane of the H$_2$S molecule. As we will see in Chapter 5, the large valence shell of third and subsequent period elements allows lone pairs to move farther apart and to push the bonding pairs closer together than in the valence shell of a smaller second-period element.

Y$_3$AE ... HX complexes

Y$_3$AE molecules such as NH$_3$, PH$_3$, and (CH$_3$)$_3$P are expected to form hydrogen-bonded complexes in which the hydrogen bond forms a tetrahedral angle with the three AY bonds. This geometry has in fact been observed for H$_3$P ... HF, as shown in Figure 4.52, and for H$_3$P ... HCl, H$_3$P ... HBr, H$_3$P ... HCN, and similar complexes with NH$_3$ and P(CH$_3$)$_3$.

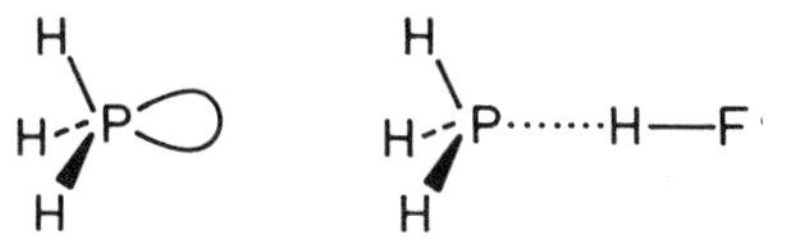

Figure 4.52 $Y_3AE...HF$ hydrogen-bonded complexes.

Hydrogen-bonded Complexes of Molecules with Multiple Bonds and Bent Bonds

A multiple bond can behave like a lone pair in forming hydrogen-bonded complexes. For example, ethene and ethyne form the T-shaped complexes with H—Cl shown in Figure 4.53. In the ethene complex the H bond is formed perpendicular to the plane of the ethene molecule because the electron density of the double bond is greatest in this direction. Similarly, the bent bonds of small-ring molecules behave as electron-pair donors in the formation of hydrogen bonds. The electron density of a bent bond is greatest in the plane of the ring, and so cyclopropane forms a hydrogen bond with HCl in this plane (Figure 4.54).

The structures of hydrogen-bonded complexes in the gas phase give direct experimental support for the arrangements of electron pairs in valence shells that are postulated by the VSEPR model.

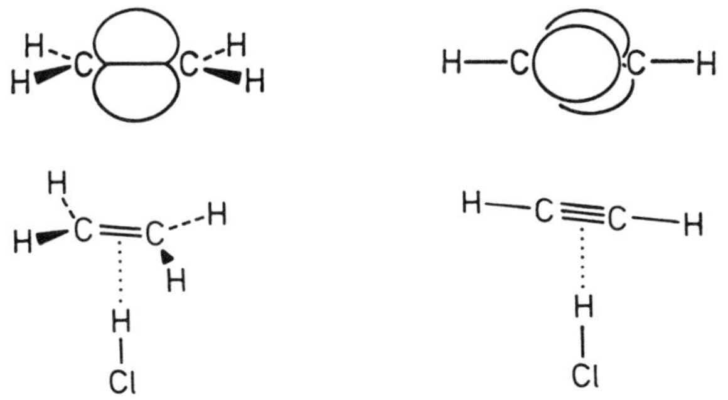

Figure 4.53 Hydrogen-bonded complexes of ethene and ethyne.

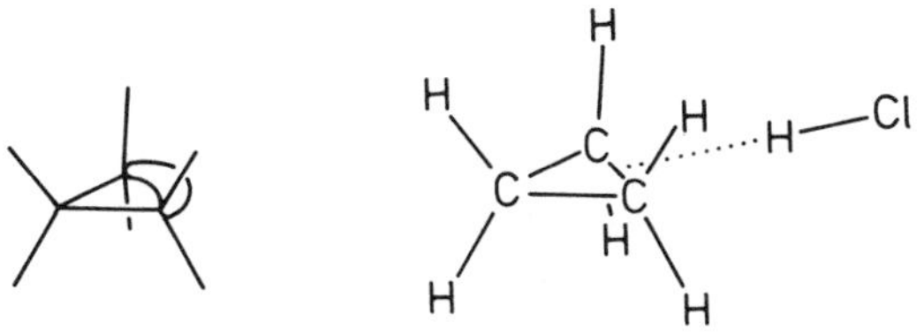

Figure 4.54 Hydrogen-bonded complex of HCl and cyclopropane.

BOND ENERGIES, BOND LENGTHS, AND MULTIPLE BONDS IN THE MOLECULES OF THE SECOND-PERIOD ELEMENTS

We pointed out at the beginning of this chapter some of the consequences for molecular geometry of the crowding of four electron pairs in the valence shells of the second-period elements, particularly nitrogen, oxygen, and fluorine. Here we take up some of the other consequences of the crowding of electron pairs, including its effect on bond energies and bond lengths and the strong tendency of these elements to form multiple bonds.

The energies of the homonuclear bonds between the elements of the second period would be expected to increase from Li to F as the increasing nuclear charge attracts the bond electrons more strongly. Table 4.13 shows that, although bond energies do increase from Li—Li to C—C, there is a sharp drop from C—C to N—N, followed by another small drop to O—O and then a slight increase to

TABLE 4.13   SINGLE-BOND ENERGIES (kJ/mol)

Li—Li	Be—Be	B—B	C—C	N—N	O—O	F—F
105	208	293	348	159	138	155
			Si—Si	P—P	S—S	Cl—Cl
			222	201	226	239
			C—H	N—H	O—H	F—H
			413	389	463	565

F—F. In contrast, in the third period this sharp decrease from group 4 to group 5 is not observed, and the Cl—Cl and S—S bond energies are slightly larger than the Si—Si bond energy. Thus the N—N, O—O, and F—F bond energies are anomalously low, and this is reflected, for example, in the reactivity of fluorine and peroxides. These unexpectedly low bond energies can be attributed to the crowding of the four electron pairs in the valence shells of nitrogen, oxygen, and fluorine and the consequent strong repulsions between the electron pairs both in the same valence shell and between the two adjacent valence shells in the molecule. These repulsions weaken the bonds between these elements, and the repulsions are particularly strong when there are unshared pairs because of the large size of their domains. The weakening of these bonds also shows up in their lengths. The F—F and O—O bonds in particular are longer than the sum of their covalent radii. Thus the F_2 molecule has a bond length of 143 pm compared to the length of 128 pm predicted from the covalent radius of fluorine. Similarly, the O—O bond length in peroxides is 147 pm compared to twice the covalent radius of oxygen, which is 132 pm.

There is no similar effect of strong electron-pair repulsions in the Si—Si, P—P, S—S, and Cl—Cl bonds because there is no crowding of the electron pairs when there are only four electron pairs in the valence shells of these elements. Bond energies also generally increase in the series C—H, N—H, O—H, F—H because hydrogen has no lone pairs, and therefore there are no lone-pair lone-pair repulsions to weaken these bonds.

One of the important properties of carbon, nitrogen, and oxygen is that they have a much greater tendency to form multiple bonds than other elements. Although in recent years compounds containing Si=Si and P=P double bonds, for example, have been synthesized (see Chapter 5), these compounds are much more reactive than compounds of the second-period elements that have double bonds unless access to the multiple bond is blocked by large substituent groups. There would appear to be two important reasons for this tendency to multiple-bond formation: (1) The electronegativities of carbon, nitrogen, and oxygen are higher than those of all other elements except the halogens and krypton, and because of their high electronegativity they can attract two or three electron pairs into the bonding region despite the repulsion between these electron pairs. (2) The strong repulsions between the electron pairs in the valence shells of carbon, nitrogen, and oxygen are decreased when two or three of these electron pairs are pulled into the bonding region. Thus multiple-bond formation by these elements is facilitated compared to third and subsequent period elements, for which the decreased valence-shell electron-pair repulsions are not important enough to compensate for the increased electron-pair repulsions arising from the crowding of two or three electron pairs into the bonding region.

The bond energies of these multiple bonds (Table 4.14) are also clearly affected by lone-pair repulsions. For carbon—carbon bonds the C=C double-bond energy is less than twice the single-bond energy, and the C≡C triple-bond energy is less than three times the single-bond energy. This is a reflection of the increased repulsion between the electron pairs when they are pulled together in the formation of a multiple bond. In contrast, the bond energy of the N=N double bond is more than twice that of the single bond, and the bond energy of the triple bond is more than five times that of a single bond. This may be attributed to the diminishing effect of lone-pair lone-pair repulsions as the distance between the lone pairs increases in this series. The N=N double bond is weaker than the C=C double bond because of lone-pair lone-pair repulsion, but the N≡N triple bond is stronger than the C≡C triple bond because the lone pairs are at 180° to each other and are therefore too far apart to repel each other strongly. The N≡N bond energy is then determined mainly by the greater electronegativity of nitrogen.

TABLE 4.14 MULTIPLE-BOND ENERGIES (kJ/mol)

C—C	N—N	C—O	C—N
348	159	356	305
C=C	N=N	C=O	C=N
602	418	799	615
C≡C	N≡N	C≡O	C≡N
835	942	1079	887

REFERENCES AND SUGGESTED READING

F. A. COTTON and G. WILKINSON, *Advanced Inorganic Chemistry*, 5th Ed., Wiley, New York, 1988.

N. N. GREENWOOD and A. EARNSHAW, *Chemistry of the Elements*, Pergamon Press, Oxford, England, 1984.

I. HARGITTAI and M. HARGITTAI, Eds., *Stereochemical Applications of Gas-Phase Electron Diffraction, Part B. Structural Information for Selected Classes of Compounds*, VCH, New York, 1988.

LANDOLT-BÖRNSTEIN, New Series, Volumes II/7 and II/15, *Structure Data of Free Poly-atomic Molecules*, Springer, Berlin, 1976 and 1987.

A. F. WELLS, *Structural Inorganic Chemistry*, 5th Ed., Oxford University Press, Oxford, England, 1984.

_5_

The Main-Group Elements
of the Third and
Subsequent Periods

The elements of the third and subsequent periods have valence shells that may contain more than four electron pairs. Although the maximum possible number of electron pairs, corresponding to the use of an s, three p, and five d orbitals, is nine, this maximum is rarely reached. For many of these elements, particularly the main-group elements of periods 3 and 4, the maximum number of electron pairs that can occupy the valence shell is six. Indeed this number is so common among the compounds of these elements that the octet rule may be supplemented by a duodecet rule, which states that the maximum number of electron pairs that can be accommodated in the valence shell of the main group elements in periods 3 and 4 is six. Nevertheless, in many of their compounds these elements have only four electron pairs in their valence shell. Thus the valence shell of a third and subsequent period element is often incomplete, and this has important consequences, some of which we have already alluded to in Chapter 4.

If the valence shell of an element of period 3 and beyond contains only four electron pairs, they are not crowded together and the electron-pair repulsions are relatively weak. A lone pair tends to spread out around the core and to push the bond pairs together until their mutual repulsion resists any further decrease in the angle between them. Because of the larger size of the valence shell of the third and subsequent period elements, two bond pairs that are at the same distance apart make a smaller angle at the nucleus in the valence shell of a third-period element than in the valence shell of a second-period element (Figure 5.1). Thus for AX_3E and AX_2E_2 molecules of these elements the bond angles may be considerably smaller than in the corresponding molecules of the second-period elements. Since the valence shell of most of these elements normally contains a maximum of six electron pairs with an octahedral arrangement, it is reasonable to suppose that the interaction between bond pairs only becomes significant as the

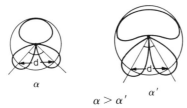

Figure 5.1 The angle between two bonding pairs at the same distance, d, apart decreases with increasing size of the valence shell.

bond angle approaches 90°, the bond-pair—bond-pair angle in an octahedron. Thus for many AX_3E and AX_2E_2 molecules of these elements the bond angles are considerably smaller than in the corresponding molecules of the second-period elements, and the bond angles in some cases approach the limiting value of 90°.

Another important consequence of the lack of crowding of four electron pairs in the valence shell of these elements is that there is little tendency for any lone pairs to delocalize into the valence shell of an adjacent atom with an incomplete valence shell. Thus, because there is no significant delocalization of lone pairs on A onto an adjacent ligand X, bond angles in AX_3E and AX_2E_2 molecules are never found to be larger than the tetrahedral angle, as they frequently are for second-period elements.

A characteristic property of the molecules of the elements of the third and subsequent periods in which there are only four electron pairs in the valence shell of the central atom is that they have a strong tendency to form additional weak secondary bonds in the solid state by attracting an electron pair of a fluorine, chlorine, or oxygen atom of another molecule in order to complete their valence shells. In contrast, except for some molecules of boron and a few gas-phase molecules, second-period elements almost always have filled valence shells in their compounds, so they do not tend to form secondary bonds in the solid state by accepting additional electron pairs. For third and subsequent period elements, secondary bond formation is quite common. We will describe several examples of solid-state structures containing secondary bonds and will show that the overall geometry including the weak secondary bonds can generally be understood in terms of the VSEPR model. In describing the geometry in such cases the ligands that are bonded by strong covalent bonds are denoted by X, and those that are bonded by weak secondary bonds are denoted by Y.

Because these elements may have more than four electron pairs in their valence shells, but rarely more than six pairs, they form many molecules whose shapes are determined by the trigonal bipyramidal arrangement of five electron pairs and the octahedral arrangement of six electron pairs, shapes that are not observed for the second-period elements.

ALKALI AND ALKALINE EARTH METALS

In the great majority of the compounds of the alkali metals the bonding is predominantly ionic, so the geometry of these compounds cannot be very usefully discussed in terms of the VSEPR model. However, the alkaline earth metals do form some molecules in which the bonding has a significant amount of covalent character, and so we discuss the geometry of these molecules in this section. These elements do not form any compounds in which they have an unshared pair in their valence shell, so they exhibit only the linear AX_2, triangular AX_3, tetrahedral AX_4, trigonal bipyramidal AX_5, and octahedral AX_6 geometries.

The vapors of the alkaline earth dihalides consist mainly of monomeric molecules in equilibrium with a small concentration of dimers. These halides have only very low vapor pressures, so their molecular structures have only been studied at high temperatures. All the beryllium and magnesium dihalides are linear, although there has been some controversy about the shape of MgF_2. All the barium dihalides are angular, while for calcium and strontium some are linear and some are angular. Table 5.1 summarizes the available data on the shapes of the alkaline earth dihalides. It can be seen that the linear form is favored for the lighter central atoms and heavier halogens, while the bent form is favored for the heavier central atoms and lighter halogens.

TABLE 5.1 BOND ANGLES (°) FOR THE GASEOUS ALKALINE EARTH DIHALIDES

MX_2 X =	F	Cl	Br	I
M = Be	180	180	180	180
Mg	155–180	180	180	180
Ca	133–155	180	173–180	180
Sr	108–135	120–143	133–180	161–180
Ba	100–115	100–127	95–135	102–105

The bent geometry of some alkaline earth metal dihalides cannot be explained by a simple application of the VSEPR model or on the basis of a simple ionic model. In using the VSEPR model we have so far assumed that the core of the central atom is spherical and nondeformable. If the outer shell of the core is completely filled, as it is for the first- and second-period elements, we expect that the core will be difficult to deform. However, for third and subsequent period elements the outer shell of the core is not completely filled because it has vacant d orbitals and it can therefore be deformed much more easily than the core of a second period element. Although the isolated ns^2np^6 core of an alkaline earth metal atom is spherical, if it is perturbed by interaction with the bonding electron pairs in the valence shell, it can be considered to consist of four somewhat localized electron pairs with a tetrahedral arrangement. The bond pairs will then tend to arrange themselves so as to minimize their interaction with these four electron-pair domains. The repulsion between the core and the bond electron pairs in a dihalide of Ca, Sr, or Ba will be a minimum when the bonding pairs are located opposite two of the faces of the tetrahedral arrangement of electron-pair domains in the outer shell of the core, thus giving a bond angle of 109.5° (Figure 5.2). If the strength of the interaction with the core is much larger than the repulsion between the ligands, the bond angle would approach 109.5°, but if the

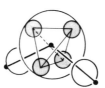

Figure 5.2 Nonlinear alkaline earth metal dihalides. The bond pairs tend to avoid the tetrahedrally arranged charge concentrations in the core of the metal atom. The greater the interaction of the bond pairs is with the core the greater is the tendency for the bond angle to decrease from 180° toward 109.5°.

bond–bond or ligand–ligand repulsion dominates, the bond angle would be expected to approach 180°. The polarizability of the core increases from Ca to Ba, so we expect the interaction between the core and the ligands to increase and the bond angle to decrease from Ca to Ba for all the halides. For the halide ligands, the polarizability decreases from iodide to fluoride and the charge density increases, so we expect the strength of the interaction with the core electrons to increase from the iodide to the fluoride and the bond angle to decrease correspondingly.

The ligand–core interaction we have described can be thought of as being similar to the interaction between the ligands and the incompletely filled d shell in many transition-metal compounds, which we discuss in Chapter 6. But because the valence shell and the outer shell of the core are not very distinctly separated in transition metals and because the outer shell of the core is very often incompletely filled, the distortion of the basic arrangement of a given number of bond pairs by a nonspherical core is much more common among the molecules of transition-metal compounds than it is among the molecules of main-group compounds.

Bis(neopentyl)magnesium is monomeric and linear in the gas phase (Figure 5.3), but most of the alkyls of these metals are dimers in the gas phase with a trigonal planar AX_3 geometry at the metal atom. In the solid state, dimethylmagnesium has an infinite chain structure with bridging methyl groups (Figure 5.4) similar to that of dimethylberyllium (Figure 4.8). The bridging alkyl groups in these compounds form three-center bonds as in dimethylberyllium, and there is a tetrahedral arrangement of four electron pairs around both carbon and

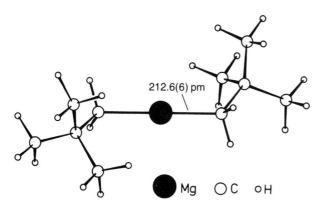

212.6(6) pm

● Mg   ○ C   ○ H

Figure 5.3 Structure of gaseous bis(neopentyl)magnesium.

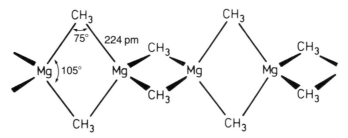

Figure 5.4   Structure of solid dimethylmagnesium.

the metal. With some bases, magnesium alkyls form tetrahedral complexes such as (quinuclidine)$_2$MgMe$_2$ (Figure 5.5).

In bis(cyclopentadienyl)magnesium the two rings are parallel to each other, but in bis(cyclopentadienyl)calcium there is a small angle between the rings (Figure 5.6). That the two rings in the calcium compound do not appear to be parallel might simply be a consequence of intramolecular motion, but it is probable that the interaction between the bonding electrons and the polarizable calcium core is responsible for the difference in the structures of the calcium and magnesium compounds, as in the case of the dihalides.

Grignard reagents, RMgX, are the most important organometallic compounds of magnesium, but their constitution in solution has been very uncertain until recently. It now seems clear that in solution there is an equilibrium between a number of different species, some of which have had their structures determined. The simplest structures are those exemplified by EtMgBr(OEt$_2$)$_2$ and PhMgBr(OEt$_2$)$_2$ which have a tetrahedral AX$_4$ geometry. There are also bridged dimers such as [EtMgBr(NEt$_3$)]$_2$ in which the magnesium has a tetrahedral AX$_4$ geometry (Figure 5.7), as well as more complex species.

An interesting example of trigonal bipyramidal AX$_5$ geometry is found in the complex ion Mg(OAsMe$_3$)$_5^{2+}$ (Figure 5.8). Calcium forms a similar complex.

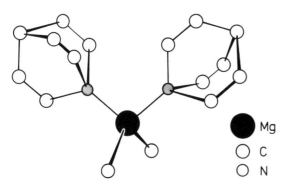

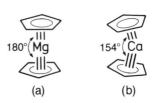

● Mg
○ C
○ N

Figure 5.5   Structure of dimethylbis(quinuclidine)magnesium, (CH$_3$)$_2$Mg. 2NC$_7$H$_{13}$ in the crystal.

Figure 5.6   (a) Linear bis(cyclopentadienyl)magnesium. (b) Angular bis(cyclopentadienyl)calcium.

(a)   (b)

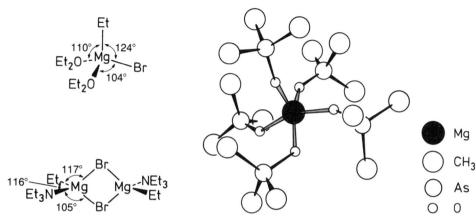

Figure 5.7 Structures of some Grignard reagents.

Figure 5.8 The trigonal bipyramidal ion $Mg(OAsMe_3)_5^{2+}$.

ALUMINUM, GALLIUM, INDIUM, AND THALLIUM

These elements have three electrons in their valence shells and form compounds in both the +3 oxidation state in which the neutral atom forms three bonds and the +1 oxidation state in which the neutral atom forms only one bond and has a lone pair. There are also a few complex ions of indium and gallium containing these elements in the +2 oxidation state. In general, the lower oxidation states become more stable with increasing atomic mass, and for thallium the +1 oxidation state is the most stable. Like the other elements discussed in this chapter, Al, Ga, In, and Tl have a strong tendency to fill their valence shells with six electron pairs and in a few compounds of In and Tl with more than six electron pairs. The observed molecular geometries are summarized in Table 5.2.

TABLE 5.2 MOLECULAR GEOMETRIES FOR ALUMINUM, GALLIUM, INDIUM, AND THALLIUM

$n + m$	Arrangement	n	m		Geometry	Example
2	Linear	2	0	AX_2	Linear	$Tl(CH_3)_2^+$
3	Trigonal planar	3	0	AX_3	Trigonal planar	$AlCl_3$
4	Tetrahedral	3	1	AX_3E	Trigonal pyramidal	$(TlOC_2H_5)_4$
		4	0	AX_4	Tetrahedral	$AlCl_4^-$
5	Trigonal bipyramidal	5	0	AX_5	Trigonal bipyramidal	$H_3Al.2N(CH_3)_3$
6	Octahedral	6	0	AX_6	Octahedral	AlF_6^{3-}

n = number of bonds, m = number of lone pairs.

AX₂ Linear Geometry

The $Tl(CH_3)_2^+$ ion has the expected linear structure in the iodide, $Tl(CH_3)_2I$.

AX₃ Trigonal Planar Geometry

The trihalides are expected to have the AX_3 triangular geometry, but most of them exist in the monomeric form only at high temperature in the gas phase in equilibrium with the dimers. The AlX_3 molecules with known structures all have the trigonal planar AX_3 geometry. The only hydride of these elements that has been studied is aluminum hydride, AlH_3. The bond lengths for those molecules that have been studied are given in Table 5.3.

TABLE 5.3   BOND LENGTHS OF AlX_3 MOLECULES

AlX_3	Al—X (pm)
AlF_3	163.1(3)
$AlCl_3$	206.8(4)
AlI_3	244.9(13)
AlH_3	171.5(10)

Trimethylgallium and trimethylthallium are monomeric in the gas phase and have the expected AX_3 geometry, while trimethylaluminum forms a mixture of monomers and dimers (see below).

AX₃E Trigonal Pyramidal Geometry

Tetrameric thallium(I) ethoxide has an interesting cubic structure in which each thallium is three coordinated and each oxygen four coordinated (Figure 5.9). The coordination around oxygen is tetrahedral, as expected, and the coordination around thallium is AX_3E trigonal pyramidal.

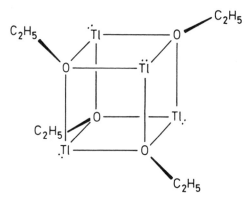

Figure 5.9   Tetrameric thallium ethoxide. Each thallium has an AX_3E trigonal pyramidal geometry.

AX$_4$ Tetrahedral Geometry

This is the most common geometry for these elements. $AlCl_3$ has an ionic structure in the solid state but exists as covalent dimeric molecules Al_2Cl_6 in the vapor state. Al_2Br_6, Al_2I_6, Ga_2Cl_6, Ga_2Br_6, Ga_2I_6, and In_2I_6 all have the same dimeric structure in the solid, liquid, and gaseous states (Figure 5.10 and Table 5.4). In each case the bridge bonds are longer than the terminal bonds. The

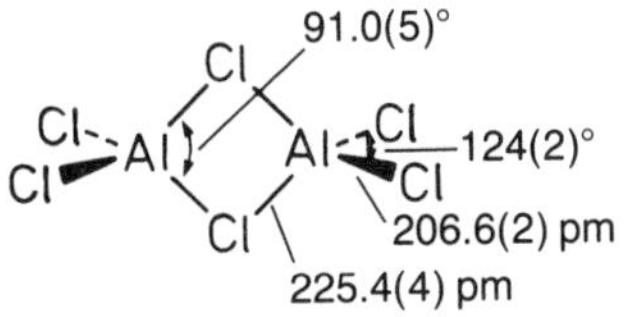

Figure 5.10 Structure of dimeric aluminum trichloride in the gas phase.

TABLE 5.4 GEOMETRICAL PARAMETERS OF GASEOUS DIMERIC ALUMINUM AND GALLIUM TRIHALIDES, M$_2$X$_6$

Molecule	M—X(t) (pm)	M—X(b) (pm)	X(t)—M—X(t) (°)	X(b)—M—X(b) (°)
Al_2F_6	160	172	124.6	80.0
Al_2Cl_6	206.6(2)	225.4(4)	123.5(16)	91.0(5)
Al_2Br_6	222.3(5)	241.7(8)	123(3)	92.2(9)
Al_2I_6	245.1(13)	264(3)	115(7)	100(4)
Ga_2Cl_6	210.0(2)	230.3(3)	125(2)	88.3(8)
Ga_2Br_6	224.6(3)	244.9(9)	128(3)	91(2)

bridging halogen has a formal positive charge, and therefore the bridge bonds are more ionic and longer and weaker than the terminal bonds. The weaker bridging bonds in Al_2Cl_6 have a stretching frequency that is 2.4 times lower than that of the terminal bonds. Although the geometry at the metal atom is approximately tetrahedral, the angles deviate considerably from the ideal angle of 109.5°. The X(b)MX(b) angle is much smaller than 109° because the two bonds are constrained in the four-membered ring, and although they are presumably bent, the angle between the bonding pairs is expected to be smaller than 109°, and so the X(t)MX(t) angle opens up to an angle that is larger than 109°. The X(b)MX(b) angle increases from the fluoride to the iodide with increasing size of the M—X(b) bond pair domain as the electronegativity of the halogen decreases. The ring is planar, but it is rather flexible and may appear puckered as a consequence of torsional vibrations.

Trimethyl- and triphenylaluminum have the same dimeric structure as the halides, but because the bridge bonds are three-center bonds the bond angle at carbon is considerably smaller than the tetrahedral angle, while that at aluminum is not much smaller than tetrahedral (Figure 5.11). In this case the bridge bonds are longer than the terminal bonds because the bridge bonds have an order of only 0.5. The structure and bonding closely resemble the structure and bonding of polymeric dimethylberyllium (Figure 4.8).

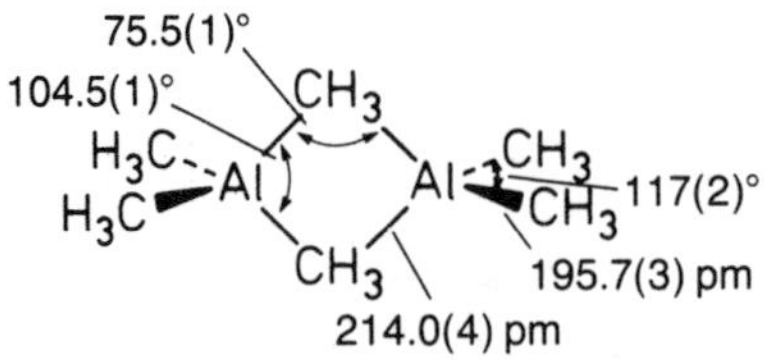

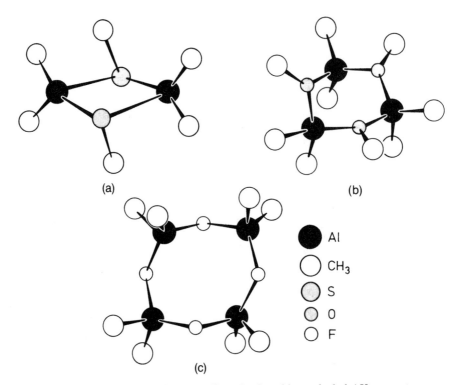

Figure 5.11 Structure of dimeric trimethyl aluminum in the gas phase.

Ring formation is quite general for compounds of the composition R_2AlX and R_2GaX, where R is an alkyl group or halogen and X may be F, Cl, OR, SR, or NR. The ring formation provides four coordination and a tetrahedral AX_4 geometry for the aluminum and gallium atoms. The rings are most often four membered, six membered, or eight membered. Several such systems are shown in Figure 5.12 (see also Figure 4.43).

(a)

(b)

● Al
○ CH_3
◐ S
◔ O
○ F

(c)

Figure 5.12 Structures of some cyclic molecules with tetrahedral AX_4 geometry at aluminum: (a) $[(CH_3)_2AlSCH_3]_2$, (b) $[(CH_3)_2AlOCH_3]_3$, (c) $[(CH_3)_2AlF]_4$.

Aluminum, gallium, and indium chlorides combine with chloride ion to give the tetrahedral complex ions $AlCl_4^-$, $GaCl_4^-$, and $InCl_4^-$. The halides, hydrides, and alkyls of these elements also form 1:1 complexes with many other donors. They all have an AX_4 approximately tetrahedral geometry. Examples include $(CH_3)_3N.AlH_3$, $(CH_3)_3N.GaH_3$, $(CH_3)_3N.Al(CH_3)_3$, $(CH_3)_3N.Ga(CH_3)_3$, $H_3N.AlCl_3$, $H_3N.GaCl_3$, $(CH_3)_3N.AlCl_3$, $H_3N.AlBr_3$, and $H_3N.GaBr_3$. Figure

5.13 compares the geometries of trimethylamine-trimethylaluminum and trimethylphosphine-trimethylaluminum. In the trimethylamine complex the CAlC angles are smaller and the Al—C bonds are longer than in the trimethylphosphine complex, which indicates that trimethylamine is a stronger donor than trimethylphosphine toward $Al(CH_3)_3$.

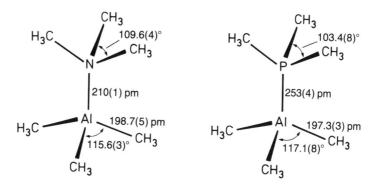

Figure 5.13   Structures of trimethylamine-trimethylaluminum and trimethylphosphine-trimethylaluminum.

Alkali metal tetrachloroaluminates and tetrafluoroaluminates have also been observed in the gas phase. They have an approximately tetrahedral $AlCl_4^-$ or AlF_4^- ion associated with the alkali metal ion by a predominantly ionic bond (Figure 5.14).

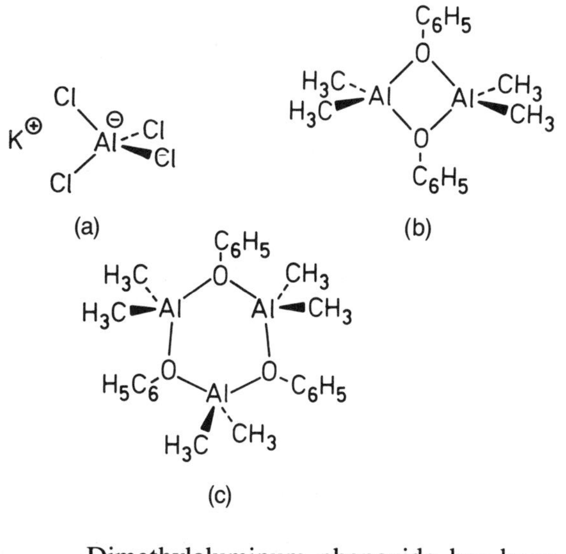

(a)

(b)

(c)

Figure 5.14   (a) Structure of potassium tetrachloroaluminate in the gas phase. (b) and (c) Structures of dimeric and trimeric dimethylaluminum phenoxide.

Dimethylaluminum phenoxide has been observed to form an equilibrium mixture of dimers and trimers (Figure 5.14) in which aluminum has an approximately tetrahedral AX_4 geometry that is less distorted in the less strained six-membered ring than in the four-membered ring.

AX$_5$ Trigonal Bipyramidal Geometry

Aluminum trihydride forms several 1:2 adducts with amines. The adduct with trimethylamine, $H_3Al.2N(CH_3)_3$, has the expected trigonal bipyramidal shape with the trimethylamine molecules in the axial positions (Figure 5.15). Because the $\overset{+}{N}(CH_3)_3$ groups are more electronegative than hydrogen, they are expected to occupy the axial positions.

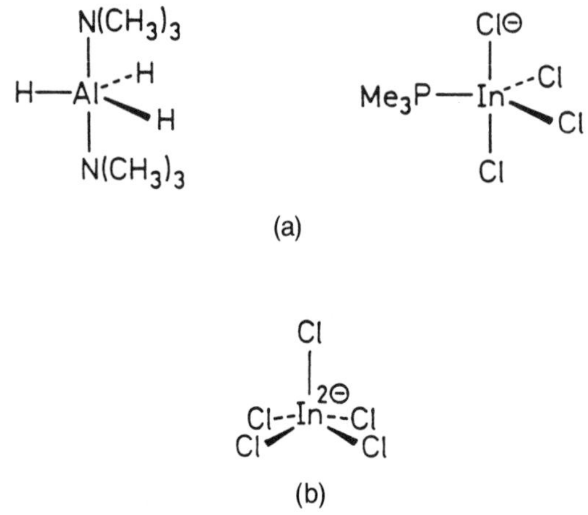

N(CH₃)₃

H—Al⁻⁻H / H

N(CH₃)₃

(a)

Cl⊖

Me₃P—In⁻⁻Cl / Cl

Cl

Cl

Cl--In--Cl 2⊖ / Cl Cl

(b)

Figure 5.15 (a) Trigonal bipyramidal AX$_5$ molecules. (b) The ion $InCl_5^{2-}$ has a square pyramidal geometry in the solid state.

The complex ion $InCl_5^{2-}$ has a square pyramidal geometry (Figure 5.15). Because the difference in energy between the trigonal bipyramid geometry and the less stable square pyramid geometry is small, it is not surprising that packing considerations and interactions between ions sometimes lead to the square pyramid being the preferred geometry in the solid state. The same structure also appears to persist in solution, but here it is probable that an approximately octahedral six-coordinated solvate, $(InCl_5 . \text{solvent})^{2-}$, is formed in which the five chlorines therefore have a square pyramid geometry. In contrast, the complex ions InX_4L^-, where X is a halogen and L is NMe_3, PMe_3, PPh_3, and Et_2O, for example, have the expected trigonal bipyramid structure with two of the halogens in the axial positions (Figure 5.15).

AX$_6$ Octahedral Geometry

These elements form many six-coordinated molecules. For example, the complex ions $Al(H_2O)_6^{3+}$, $Al(OH)_6^{3-}$, and AlF_6^{3-} are all octahedral. Complex fluorides such as Tl_2AlF_5 and $KAlF_4$ in the crystal also contain octahedral AlF_6 groups sharing two corners to give the linear polymeric $(AlF_5^{2-})_n$ anion (Figure 5.16) and sharing four corners to give the planar polymeric cation $(AlF_4^-)_n$ (Figure 5.16), respectively. The binuclear complex anion $Tl_2Cl_9^{3-}$ has a structure in which two octahedral $TlCl_6$ groups share a face (Figure 5.16).

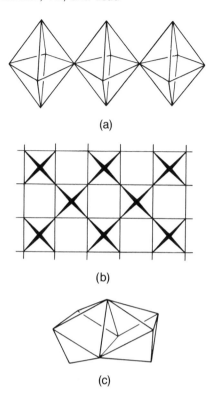

(a)

(b)

(c)

Figure 5.16 (a) Linear polymeric structure of the $(AlF_5^{2-})_n$ anion. (b) Planar polymeric structure of the $(AlF_4^-)_n$ anion. (c) Dimeric structure of the $Tl_2Cl_9^{3-}$ ion.

SILICON, GERMANIUM, TIN, AND LEAD

These elements have four electrons in their valence shells, and they can use all four to form bonds giving the +4 oxidation state, or they may use only two, leaving one unshared pair and giving the +2 oxidation state. The observed molecular geometries are summarized in Table 5.5.

TABLE 5.5 MOLECULAR GEOMETRIES FOR SILICON, GERMANIUM, TIN, AND LEAD

$n + m$	Arrangement	n	m		Geometry	Example
3	Trigonal	2	1	AX_2E	Bent	$SiCl_2$
		3	0	AX_3	Trigonal planar	$H_2C{=}Si(CH_3)_2$
4	Tetrahedral	3	1	AX_3E	Trigonal pyramidal	$SnCl_3^-$
		4	0	AX_4	Tetrahedral	$SiCl_4$
5	Trigonal bipyramidal	4	1	AX_4E	Disphenoidal	GeF_4 in crystal
		5	0	AX_5	Trigonal bipyramidal	$[(CH_3)_2NSiH_3]_5$
6	Octahedral	6	0	AX_6	Octahedral	SnF_6^{2-}

n = number of bonds, m = number of lone pairs.

AX₂E Angular Geometry

The group 4 dihalides are angular AX_2E molecules with a bond angle of less than
120°. Two trends in the observed bond angles can be seen in the data given in
Table 5.6: (1) The bond angles increase in the order $F < Cl < Br < I$ for a given
central atom. This order can be attributed to increasing bond–bond repulsions

TABLE 5.6 BOND ANGLES (°) OF GROUP 4 DIHALIDES

AX_2 $X=$	F	Cl	Br	I
A = C	104.8	104.9	—	—
Si	100.8	103.1(6)	102.9(3)	—
Ge	97.2	100.4(4)	101.4(9)	—
Sn	—	99(1)	100.0(7)	103.8(7)
Pb	98(2)	98(2)	99(2)	99.7(8)

with decreasing ligand electronegativity and therefore increasing size of the bond
pairs. (2) For the same halide, the bond angles decrease in the order $C > Si >
Ge > Sn > Pb$. This order can be attributed to decreasing bond–bond repulsions
with increasing size of the central atom. Another example of a group 4 angular
AX_2E molecule is germanium bis(hexamethyl-disilyl)methane (Figure 5.17),
which has a bond angle at germanium of 107(2)°.

Germanocene, $(C_5H_5)_2Ge$, that is, bis(cyclopentadienyl)germanium, and its
tin and lead analogues are all AX_2E molecules (Figure 5.18). In the gas phase
they are bent with angles at the central atom of about 130°, which is consistent
with our earlier formulation of the metal cyclopentadienyl bond as a type of triple
bond, because we expect the large size of the triple-bond domains to cause the
bond angles to be larger than the ideal angle of 120°. However, in molecules in
which the cyclopentadienyl rings have bulky substituents such as decaphenyl
stannocene $[(C_6H_5C)_5]_2Sn$, the rings are parallel.

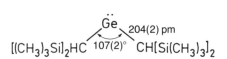

Figure 5.17 Angular AX_2E structure of
germanium bis(hexamethyldisilyl)methane.

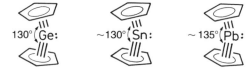

Figure 5.18 Angular structure of bis(cyclopentadienyl)germanium and its tin and lead analogues.

AX₃ Trigonal Planar Geometry

The first compound to be prepared with a silicon carbon double bond,
1,1-dimethylsilaethene, $H_2C=Si(CH_3)_2$ (Figure 5.19), is a short-lived species
obtained by pyrolysis of 1,1-dimethylsilacyclobutane. As expected it has =Si—
angles greater than 120°, and similar bond angles have been obtained for

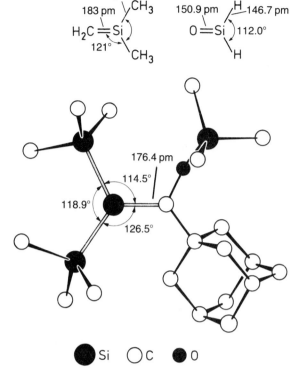

191 pm
183 pm 150.9 pm 146.7 pm

Figure 5.19 Planar AX$_3$ structures of H$_2$CSi(CH$_3$)$_2$ (observed) and OSiH$_2$ (calculated).

176.4 pm
114.5°
118.9°
126.5°

● Si ○ C ◐ O

Figure 5.20 Structure of [(CH$_3$)$_3$Si]$_2$Si=C(C$_{10}$H$_{15}$)[OSi(CH$_3$)$_3$] with a planar AX$_3$ geometry at the central silicon.

H$_2$Si=O, H$_2$Si=S, and H$_2$Si=PH by ab initio calculations (Figure 5.19). The crystal molecular structure of a silaethene derivative is shown in Figure 5.20. Again, the silicon has trigonal planar geometry, and the SiC bond has a length of 176.4 pm, which is close to the value of 174 pm predicted from the double-bond radii of carbon and silicon and much shorter than the sum of the single-bond radii, which is 194 pm.

AX$_3$E Trigonal Pyramidal Geometry

The dihalides of germanium and tin have polymeric structures in the solid state in which the tin and germanium atoms have a pyramidal AX$_3$E geometry. In the fluoride SnF$_2$, cyclic tetramers are held together by fluorine bridges so that the overall geometry around each tin atom is approximately square pyramidal AX$_3$Y$_2$E (Figure 5.21). In the chloride SnCl$_2$, there are infinite chains formed by triangular pyramidal SnCl$_3$ units in which two of the chlorines form bridges to neighboring tin atoms (Figure 5.22). As is always the case, these bridge bonds are longer than the bonds to terminal (nonbridging) chlorines.

Germanium difluoride also has a fluorine-bridged chain structure in which each germanium has an AX$_3$E pyramidal geometry; but if an additional weak fluorine bridge is counted, then there is a distorted disphenoidal, AX$_3$YE, geometry around germanium (Figure 5.23). The axial bonds are unsymmetrical

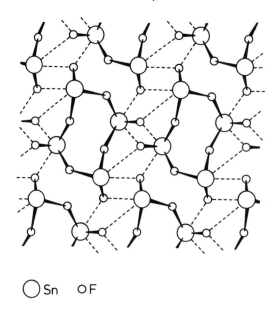

○ Sn   ○ F

Figure 5.21   Structure of crystalline SnF$_2$, which consists of cyclic tetramers in which each tin has a pyramidal AX$_3$E geometry that are held together by fluorine bridges, giving each tin an overall square pyramidal AX$_3$Y$_2$E geometry.

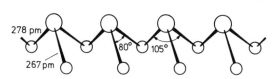

278 pm

80° 105°

267 pm

Figure 5.22   Infinite chain structure of SnCl$_2$ in which each tin has a pyramidal AX$_3$E geometry.

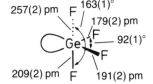

257(2) pm   F   163(1)°
                179(2) pm
            F
      Ge   92(1)°
                F
209(2) pm   F   191(2) pm

Figure 5.23   Germanium difluoride has an approximate disphenoidal AX$_3$YE geometry in the solid state in which each germanium forms three strong bonds and one rather weak bridge bond.

and considerably longer than the equatorial bonds, and the angle between the axial bonds is 163° and that between the equatorial bonds is only 92°.

The ion SnCl$_3^-$ has a pyramidal AX$_3$E geometry at tin (Figure 5.24). There are also three secondary bonds (chlorine bridges) that complete an overall AX$_3$Y$_3$E distorted octahedral geometry around the tin atom.

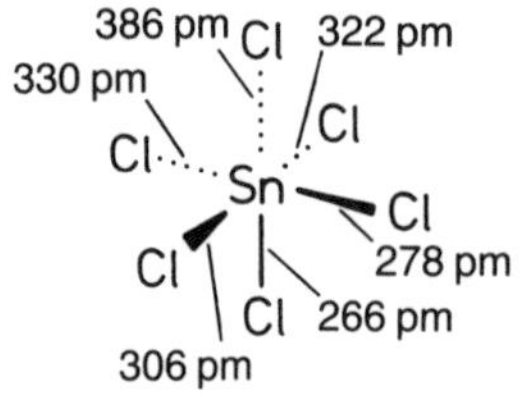

386 pm Cl 322 pm
330 pm   Cl
Cl   Sn   Cl
Cl   278 pm
     Cl 266 pm
306 pm

Figure 5.24   Geometry of tin in the SnCl$_3^-$ ion in the solid state.

AX₄ Tetrahedral Geometry

Simple molecules of the type AX_4, such as $SiCl_4$, $Ge(CH_3)_4$, $SnCl_4$, and $Pb(CH_3)_4$, have the expected tetrahedral structure, and molecules with two or more different ligands, such as $Sn(CH_3)_2Cl_2$, show the expected small deviations from the regular tetrahedral structure. The same tetrahedral arrangement of four bonds is found in the various forms of silica, the silicates, and the siloxanes. For example, the β-cristobalite form of silica has the structure shown in Figure 5.25,

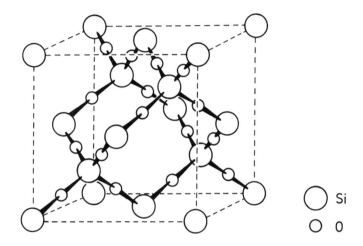

Si ⃝

O ⃝

Figure 5.25   Crystal structure of the β-cristobalite form of silica.

with tetrahedral bonds around silicon and a rather large bond angle of 150° at oxygen. The bond angle at an oxygen atom between two silicon atoms is always larger than the tetrahedral angle and it varies rather widely from compound to compound. These bond angles at oxygen have been discussed in Chapter 4.

Silicon sulfide, SiS_2, consists of tetrahedral SiS_4 units sharing opposite edges to form linear chains (Figure 5.26). Germanium sulfide, GeS_2, has a similar structure. The cyclic $[Si(CH_3)_2]_2S_2$ molecule has a closely related structure (Figure 5.26). In these and other molecules the bond angle at a sulfur atom situated between two silicon atoms is smaller than the tetrahedral angle, whereas

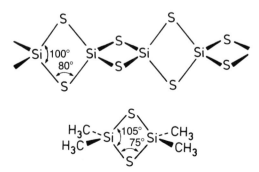

Figure 5.26   Structure of polymeric silicon sulfide and dimeric dimethylsilicon sulfide.

the bond angle at an oxygen situated between two silicon atoms is always greater
than the tetrahedral angle. The strong repulsions between the oxygen lone pairs
cause them to delocalize into the valence shells of the adjacent silicon atoms, but
because the repulsions between the sulfur lone pairs are much weaker there is
little tendency for them to delocalize.

Comparison of the tetrahalides and dihalides of the same central atom
(Table 5.7) shows not only a decrease in the bond angle XAX on going from the
AX_4 to the AX_2E geometry, but also that this bond angle decrease is accompanied by a lengthening of the AX bond. This bond lengthening is paralleled by
an appreciable decrease in the bond stretching force constant.

It is of interest to consider the effect on the bond angles of successively
substituting the ligands in an SiX_4 molecule with a different ligand (Table 5.8).
For example, on substituting a hydrogen in SiH_4 with a more electronegative
fluorine to obtain $FSiH_3$, there is an increase in the HSiH angles. Likewise,
starting with SiF_4 and going to $HSiF_3$ leads to a decrease in the FSiF angles.
Furthermore, the F—Si—F angle in F_3SiH is smaller than the H—Si—H angle in
H_3SiF. On substituting a second hydrogen in SiH_4 by fluorine there is a further
increase in the HSiH angle. Likewise, on going from $HSiF_3$ to H_2SiF_2 there is a
further decrease in the FSiF angle.

The effect of substituting one F in SiF_4 and of substituting one H in SiH_4 can
be unambiguously predicted because there are only two kinds of bond–bond

TABLE 5.7 BOND ANGLES, (°), BOND LENGTHS, r (pm), AND BOND STRETCH-
ING FORCE CONSTANTS, f (N/m), OF AX_2E_2 AND AX_4 MOLECULES

Molecule	Angle	r	f	Molecule	Angle	r	f
SiF_2	100.8	159.0	503	SiF_4	109.5	155.3(2)	657
$SiCl_2$	103.1(6)	208.9(4)	229	$SiCl_4$	109.5	202.0(4)	337
$SiBr_2$	102.9(3)	224.9(5)	–	$SiBr_4$	109.5	—	—
GeF_2	97.2	173.2	408	GeF_4	109.5	167(3)	557
$GeCl_2$	100.4(4)	218.6(4)	206	$GeCl_4$	109.5	211.3(3)	280
$GeBr_2$	101.4(9)	233.7(13)	162	$GeBr_4$	109.5	227.2(3)	241
$SnCl_2$	99(1)	234.6(7)	210	$SnCl_4$	109.5	228.1(4)	263
$PbCl_2$	98(2)	244.5(5)	170	$PbCl_4$	109.5	243(4)	210

TABLE 5.8 BOND ANGLES (°) AND BOND LENGTHS (pm) IN
FLUOROSILANE MOLECULES

Angles	SiH_4	$FSiH_3$	F_2SiH_2	F_3SiH	SiF_4
HSiH	109.5	110.5	114.2(3)	—	—
HSiF	—	108.4(5)	108.7	110.6(1)	—
FSiF	—	—	107.7(2)	108.3	109.5
Bonds	SiH_4	$FSiH_3$	F_2SiH_2	F_3SiH	SiF_4
SiH	148.1(6)	147.0(5)	146.5(3)	144.7(1)	—
SiF	—	159.1(2)	157.6(2)	156.2(1)	155.4(1)

interactions and only two bond angles, one of which is determined by the other by virtue of the threefold molecular symmetry. But it is not possible to make such an unambiguous prediction of the consequences of the second substitution. In F_2SiH_2, there are three different kinds of bond–bond interactions, one SiF/SiF, one SiH/SiH, and four SiF/SiH, and there are three different bond angles, of which two must be known to determine the third. Thus it is not possible to predict rigorously the bond-angle changes from $FSiH_3$ to F_2SiH_2, nor from F_3SiH to F_2SiH_2. Nevertheless, all the observed bond-angle changes are consistent with the VSEPR model. For example, the three different bond angles in the F_2SiH_2 molecule increase in the order FSiF, FSiH, HSiH, as expected on the basis of the greater electronegativity of F than H. The SiH and SiF bond lengths are also given in Table 5.8. They show the expected trends. The Si—H bonds shorten with increasing fluorine substitution, whereas the Si—F bonds lengthen from SiF_4 to H_3SiF.

Analogous bond-angle variations are observed in the series $(CH_3)_4Si$, $(CH_3)_3SiF, \ldots , SiF_4$. The data are presented in Table 5.9.

There are many other examples of tetrahedral AX_4 geometry at silicon. For example, disilane and its derivatives X_3SiSiX_3 have nearly ideal tetrahedral XSiX angles (Figure 5.27). For X = H, Cl, F, or CH_3, these angles are $108.6(4)°$,

TABLE 5.9 BOND ANGLES (°) AND BOND LENGTHS (pm) IN METHYLFLUOROSILANE MOLECULES

Angles	$(CH_3)_4Si$	$(CH_3)_3SiF$	$(CH_3)_2SiF_2$	CH_3SiF_3	SiF_4
CSiC	109.5	111.5(2)	116.7(6)	—	—
CSiF	—	107.4	108.7	112.0	—
FSiF	—	—	104.6(4)	106.8(5)	109.5
Bonds	$(CH_3)_4Si$	$(CH_3)_3SiF$	$(CH_3)_2SiF_2$	CH_3SiF_3	SiF_4
SiC	187.6(2)	184.8(1)	183.6(1)	182.8(4)	—
SiF	—	160.0(1)	158.6(1)	157.0(1)	155.4(1)

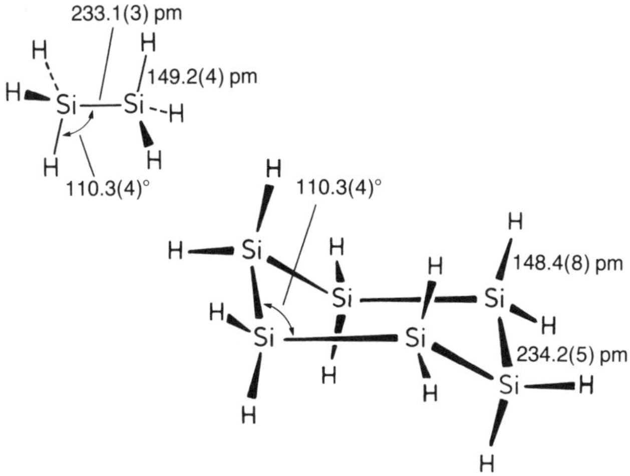

Figure 5.27 Structures of disilane and cyclohexasilane.

109.7(6)°, 108.6(3)°, and 110.5(4)°, respectively. In cyclohexasilane the SiSiSi bond angle is 110.3(4)°; the molecule has a chair conformation (Figure 5.27).

AX₄E Dlsphenoidal Geometry

We have seen that germanium difluoride (Figure 5.23) has a chain structure in the solid state in which germanium has an AX_3E pyramidal geometry if only the three closest fluorine atoms are considered. However, if an additional fluorine bridge is taken into account, the germanium has an approximately disphenoidal AX_4E geometry with the lone pair in the equatorial position of a trigonal bipyramid.

In PbO and SnO (Figure 5.28) the metal atom has a square pyramidal AX_4E geometry with the lone pair presumably in the apical position rather than the expected disphenoidal geometry. However, because the difference in energy between the square pyramidal and the trigonal bipyramidal arrangements of five electron pairs is small, it is quite possible that the square pyramid structure could be favored in an extended structure in the solid state.

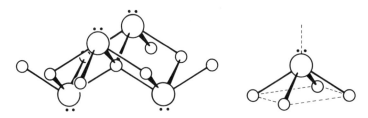

Figure 5.28   One layer of the solid-state structure of SnO_2 and PbO_2 in which each metal atom has a square pyramidal AX_4E geometry.

AX₅ Trigonal Bipyramidal Geometry

The ions $SnCl_5^-$ and $(CH_3)_2SnCl_3^-$ have the AX_5 trigonal bipyramidal geometry. Several trimethyltin compounds such as $(CH)_3SnClO_4$, $(CH_3)_3SnBF_4$, and $(CH_3)_3SnF$ have polymeric structures in which planar trimethyltin groups are linked by bridging groups or atoms, giving an overall trigonal bipyramidal coordination around tin (Figure 5.29).

In 1-methylsilatrane and 1-fluorosilatrane (Figure 5.30) there appears to be a nitrogen–silicon bond in the crystal, whereas in the gas the N—Si separation is very large, suggesting a much weaker interaction. Perhaps intermolecular interactions compress the molecule in the crystal to facilitate the formation of an N—Si bond. The NSi distances and silicon bond angles are given in Table 5.10 for both phases. The gas-phase geometry may be considered to be distorted tetrahedral, while the crystal-phase structure is distorted trigonal bipyramidal.

Another molecule with trigonal bipyramidal AX_5 geometry at silicon is the cyclic pentamer of N,N-dimethyl(silyl)amine, $[(CH_3)_2NSiH_3]_5$. In this ring structure, five-coordinate silicon is linked to four-coordinate nitrogen (Figure 5.31).

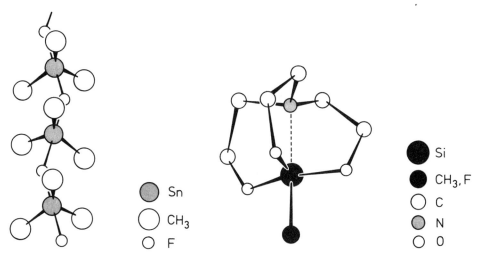

Figure 5.29   Structure of $(CH_3)_3SnF$ in solid state.

Figure 5.30   1-Methylsilatrane and related molecules.

TABLE 5.10   NSi DISTANCES (pm) AND SILICON BOND ANGLES (°) OF 1-METHYLSILATRANE AND 1-FLUOROSILATRANE IN THE GAS AND IN THE CRYSTAL

Parameter	1-Methylsilatrane		1-Fluorosilatrane	
	Gas	Crystal	Gas	Crystal
NSi	232(1)	204.2(1)	245(5)	217.5(4)
NSiO	79(2)	82.7(2)[a]	81.3(3)	86.0(1)[a]
OSiO	116.5(9)	118.3(4)[a]	117.8(1)	119.6(2)[a]
CSiO/FSiO	101(2)	97.3(5)[a]	98.7(3)	93.8(2)[a]

[a]Mean angles in the crystal.

All the Si—N bonds in the ring are equivalent. The hydrogens are in the equatorial plane and the nitrogens in the axial positions of the trigonal bipyramid geometry. The SiN bond length of 195 pm is somewhat larger than the sum of the single-bond radii, which is 187 pm.

A derivative of $(CH_3)_2NSiH_3$, chlorosilyl-N,N-dimethylamine, $ClH_2SiN(CH_3)_2$, is monomeric in the gas phase with a nearly planar nitrogen, a tetrahedral silicon, and an Si═N double bond (Figure 5.32). In the crystal, it forms dimers (Figure 5.32) with nonequivalent Si—N bonds of lengths 181 and 205 pm. The silicon has an AX_5 trigonal bipyramidal geometry, with the short SiN bond equatorial and the long SiN bond axial. Chlorine occupies the remaining axial position.

In some cases the ligand geometry may dictate a square pyramidal rather than the trigonal bipyramidal geometry at silicon. This is the case for some chelating ligands (Figure 5.33).

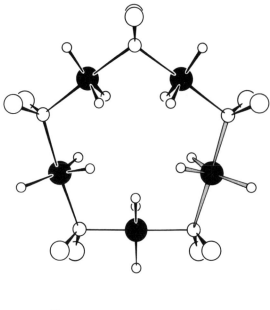

●Si   ○CH₃   ○N   OH

Figure 5.31 Cyclic pentamer of *N,N*-dimethylsilylamine.

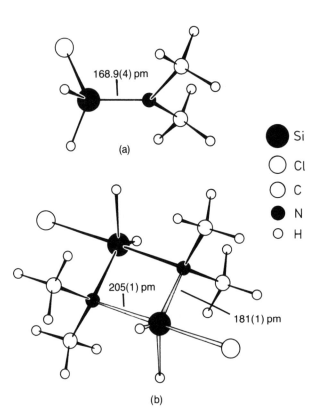

● Si

○ Cl

○ C

● N

○ H

Figure 5.32 Structure of chorosilyl-*N,N*-dimethylamine. (a) Gas phase. (b) Solid state.

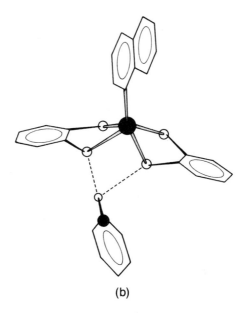

Figure 5.33 (a) Anions with trigonal bipyramidal AX_5 geometry at silicon. (b) $C_{10}H_7Si(O_2C_6H_4)_2 \cdot C_5H_5NH$ that has a square pyramidal AX_5 geometry at silicon.

AX$_6$ Octahedral Geometry

Numerous six-coordinated complexes are known, such as SnF_6^{2-}, $Pb(OH)_6^{2-}$, $SnCl_4(acetone)_2$, $SnCl_4(OPCl_3)_2$, and $GeCl_4(pyridine)_2$. They all have the expected AX$_6$ octahedral geometry. This same geometry is also found in one form of GeO_2 and in SnO_2, which have the rutile structure (Figure 5.34). These are generally described as ionic crystals, but it is reasonable to suppose that the bonds have a certain amount of covalent character. SnF_4 has a polymeric structure like AlF_4^- (Figure 5.16) with octahedral geometry around the tin. SnS_2 has a layer structure in which each tin forms six octahedral bonds and each sulfur has three pyramidal bonds and one lone pair in the fourth tetrahedral position (Figure 5.35).

Six-coordinate silicon occurs in a few silicates, although the vast majority contain four-coordinate silicon. A very high pressure form of silica called stishovite has the rutile structure (Figure 5.34) in which the SiO$_6$ geometry is only slightly distorted from an ideal octahedron.

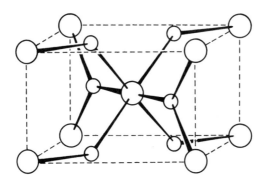

◯ Si, Ge, Sn ◯ O

Figure 5.34 Rutile structure of SnO_2, one form of GeO_2, and the very high pressure form of SiO_2 called stishovite.

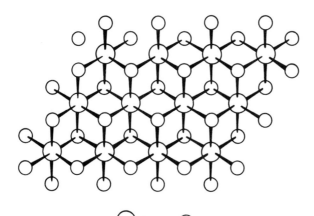

◯ Sn   ◯ S

Figure 5.35 Layer structure of SnS_2.

PHOSPHORUS, ARSENIC, ANTIMONY, AND BISMUTH

These elements form compounds either using all their electrons to form bonds, as in the +5 oxidation state, or three of their electrons, leaving one unshared pair, as in the +3 oxidation state. The observed molecular geometries are summarized in Table 5.11.

TABLE 5.11 MOLECULAR GEOMETRIES FOR PHOSPHORUS, ARSENIC, ANTIMONY, AND BISMUTH

$n + m$	Arrangement	n	m		Geometry	Example
3	Trigonal	2	1	AX_2E	Bent	F_3C—P=CF_2
		3	0	AX_3	Trigonal planar	$(CH_3)_2C$=$P(CH_3)Cr(CO)_5$
4	Tetrahedral	3	1	AX_3E	Trigonal pyramidal	PCl_3
		4	0	AX_4	Tetrahedral	$OPCl_3$
5	Trigonal bipyramidal	4	1	AX_4E	Disphenoidal	$SbCl_3.NH_2C_6H_5$
		5	0	AX_5	Trigonal bipyramidal	PF_5
6	Octahedral	5	1	AX_5E	Square pyramidal	$SbCl_5^{2-}$
		6	0	AX_6	Octahedral	PF_6^-

n = number of bonds, m = number of lone pairs.

AX_2E Angular Geometry

The molecule F_3C—P=CF_2 has an angular AX_2E geometry at phosphorus with a bond angle of 108.8(8)° (Figure 5.36). It is rather unstable in the gas phase and dimerizes in the crystal (Figure 5.40). Other molecules of phosphorus with this geometry include HP=CH_2, $P(NMe_2)_2^+$, and cyclo-C_5H_5P (Figure 5.36).

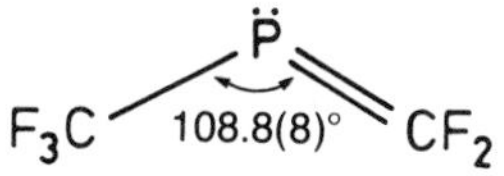

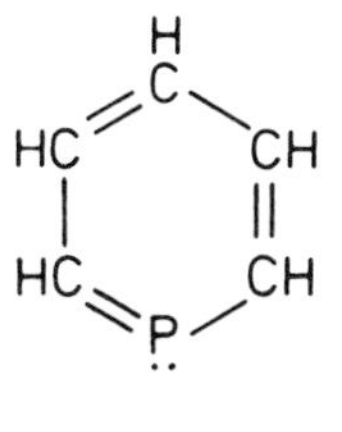

Figure 5.36 Phosphorus has an angular AX_2E geometry in F_3CP=CF_2 and cyclo-C_5H_5P.

AX_2E_2 Angular Geometry

This type of geometry is found in the anions $P(CN)_2^-$ and P_7^{3-} (Figure 5.37).

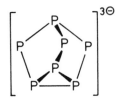

Figure 5.37 Cage structure of P_7^{3-}.

AX₃ Trigonal Planar Geometry

In the crystal the molecule shown in Figure 5.38 has a trigonal planar geometry at phosphorus. The angles involving the P=C double bond are larger than the angle between the single bonds.

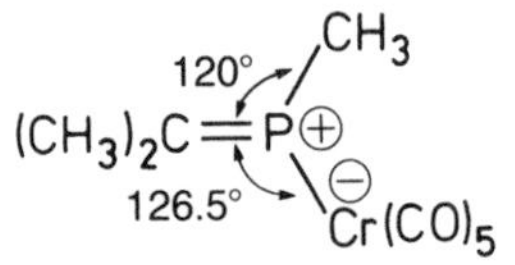

$$(CH_3)_2C \overset{120°}{=} \overset{/CH_3}{\underset{126.5°}{P\oplus}} \underset{Cr(CO)_5}{\overset{\ominus}{\diagdown}}$$

Figure 5.38 Molecule with planar AX₃ geometry at phosphorus.

AX₃E Trigonal Pyramidal Geometry

Bond angles for the trihalides, trihydrides, and some related molecules are given in Table 5.12. The lone pair causes the bond angles to be less than the ideal tetrahedral angle in every case. The bond angles decrease from left to right in the

TABLE 5.12 BOND ANGLES (°) FOR AX₃E MOLECULES

X	PX₃	AsX₃	SbX₃
F	97.7(1)	95.8(1)	—
Cl	100.3(1)	98.9(2)	97.1(2)
Br	101.0(4)	99.8(2)	98.2(6)
I	102	100.2(4)	99(1)
H	93.2(1)	92.1(1)	91.6(1)
CH₃	98.6(2)	96.1(5)	94.1(5)
SiH₃	96.8(5)	94.1(2)	88.6(2)

table as the electronegativity of the central atom decreases, and they increase from the fluoride to the iodide with decreasing electronegativity of the ligand. However, as we have seen in several other cases, the smallest bond angles are observed for the hydrides. Their angles also decrease in the expected manner with decreasing electronegativity of the central atom from phosphorus to antimony. The HPF angle in HPF₂ is intermediate, 96.3(5)°, between HPH of PH₃ and FPF of PF₃. Similarly, FPCl of ClPF₂ is 99(1)°, in between ClPCl of PCl₃ and FPF of PF₃.

P(SiH₃)₃ and its analogs are pyramidal, unlike N(SiH₃)₃, which is planar. In contrast to nitrogen, in these molecules phosphorus, arsenic, and antimony have incomplete valence shells in which the electron-pair repulsions are relatively weak compared to those in the second-period elements, and therefore they have little tendency to delocalize their lone pairs.

P_2F_4 and P_2I_4 have the centrosymmetric *anti* structure shown in Figure 5.39 in the solid state, with a pyramidal AX_3E geometry around each P atom. The FPF angle is 99° and the IPI angle is 102°.

Both of the phosphorus atoms in the dimer of $F_3C\!-\!P\!=\!CF_2$ have a trigonal pyramidal geometry. The four-membered ring is considerably puckered (Figure 5.40). As expected, the bond angle at carbon is larger than that at phosphorus.

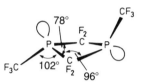

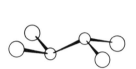

Figure 5.39 Structure of P_2F_4 and P_2I_4.

Figure 5.40 Structure of the dimer of $F_3CP\!=\!CF_2$ in which each phosphorus has a pyramidal AX_3E geometry.

The tetrahedral P_4 and As_4 molecules have an AX_3E geometry around each atom (Figure 5.41). Although the bond angles are only 60°, it is reasonable to presume that there is considerable bond bending due to bond–bond repulsions, and that the maximum of the electron density in each bond lies outside the internuclear axis (Figure 5.41), so the angles between the bonding pairs are considerably larger than 60°.

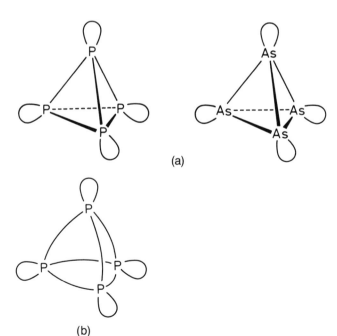

(a)

(b)

Figure 5.41 (a) Tetraphosphorus and tetraarsenic. (b) Bent bonds in P_4.

Other molecules in which the geometry is of the AX_3E type include the oxides of the +3 oxidation state of these elements. The molecules P_4O_6, As_4O_6, and Sb_4O_6 have the structure shown in Figure 5.42. The bond lengths and bond

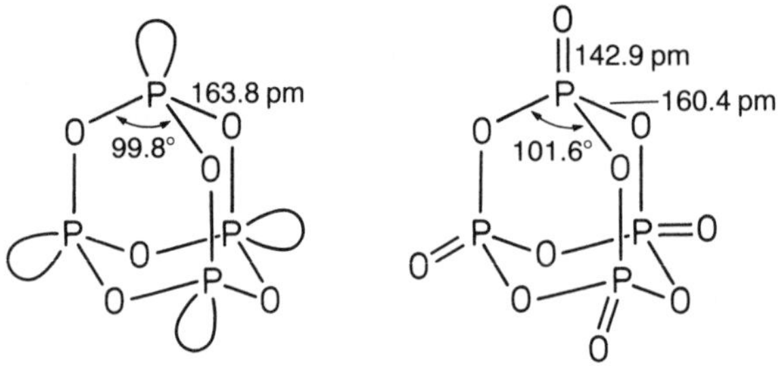

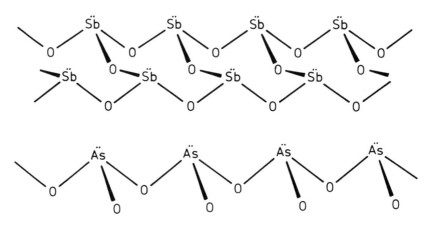

Figure 5.42 Structures of P_4O_6 and P_4O_{10}.

angles are given in Table 5.13. Antimony(III) oxide has another form with a double-chain structure (Figure 5.43). The polyarsenite ion, $(AsO_2^-)_n$, has a similar single chain of pyramidal AsO_3 groups (Figure 5.43).

TABLE 5.13 BOND LENGTHS AND BOND ANGLES OF A_4O_6 MOLECULES

Compound	A—O (pm)	AOA (°)
P_4O_6	163.8(3)	126.4(7)
As_4O_6	178(2)	128(2)
Sb_4O_6	200(2)	129(3)
$(OP)_4O_6$	160.4(5)	123.5(7)

Figure 5.43 Chain structures of antimony(III) oxide and the arsenite ion $(AsO_2^-)_n$.

Although the sulfides of phosphorus(III) and arsenic(III) often have different structures from the oxides, in each case the phosphorus and arsenic atoms have the pyramidal AX_3E geometry (Figure 5.44).

Figure 5.44 Structures of P_4S_3 and As_4S_4.

AX_4 Tetrahedral Geometry

There are numerous molecules of this type, such as PX_4^+, $O{=}PX_3$, and $S{=}PX_3$. They all have the expected tetrahedral shape. When the ligands are identical, as in PCl_4^+, $As(C_2H_5)_4^+$, and PO_4^{3-}, the tetrahedra are regular. In molecules of the type POX_3, the XPX angles are always found to be less than 109.5° as a consequence of the greater repulsion exerted by a double bond than by a single bond (see Figure 5.45), but the XPX bond angles in these molecules are invariably greater than in the PX_3E molecules. Electron withdrawal by the electronegative oxygen atom will cause the PX bonds to be less polar in OPX_3 than in PX_3E and therefore they will repel each other more strongly, giving a larger bond angle. These bond angle differences are also consistent with the lone-pair domain being larger and therefore exerting a greater repulsion toward the other bonds than a P=O or P=S double bond. Similar bond angles are found for the analogous molecules with arsenic as the central atom (Figure 5.46).

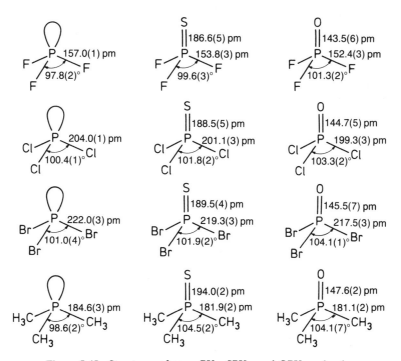

Figure 5.45 Structures of some PX_3, SPX_3, and OPX_3 molecules.

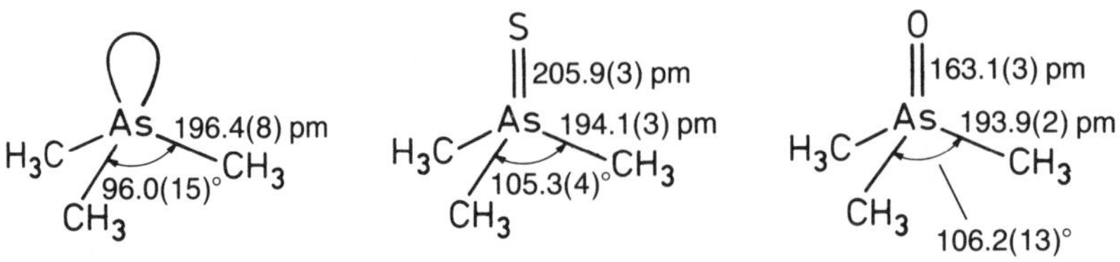

Figure 5.46 Structures of $As(CH_3)_3$, $SAs(CH_3)_3$, and $OAs(CH_3)_3$.

The bond lengths in OPX_3, SPX_3, and EPX_3 molecules are of interest. With increasing electronegativity of the X ligand, there is a marked shortening of the P=S as well as of the P=O bond. For any X ligand, the PX bond of the PX_3E molecule is the longest. Analogous observations can be made for the As—C bonds in $AsO(CH_3)_3$, $AsS(CH_3)_3$, and $As(CH_3)_3$.

Phosphoric oxide, P_4O_{10}, also contains tetrahedrally coordinated phosphorus. Each O_3PO group can be compared with the O_3PE group of P_4O_6 (Figure 5.42). We see again that the lone pair of phosphorus repels the neighboring bonds more strongly than the P=O double bond in analogous situations.

AX_4E Disphenoidal Geometry

The molecule $SbCl_3.C_6H_5NH_2$ and the anion $Sb(SCH_2CO_2)_2^{2-}$ have the expected trigonal bipyramidal arrangement of the two axial ligands, two equatorial ligands, and an equatorial lone pair around antimony (Figure 5.47). In one crystalline form of SbO_2 there are six-coordinated Sb(V) atoms and four-coordinated Sb(III) atoms. The latter have the expected disphenoidal geometry (Figure 5.47). Although there are several complex ions with the general formula MX_4^-, most of them are polymerized and only a few, such as PBr_4^- and SbF_4^-, are known in the monomeric form and have the AX_4E disphenoidal geometry (Figure 5.47). In the anion $Sb_2F_7^-$ both antimony atoms have an AX_4E disphenoidal geometry, with

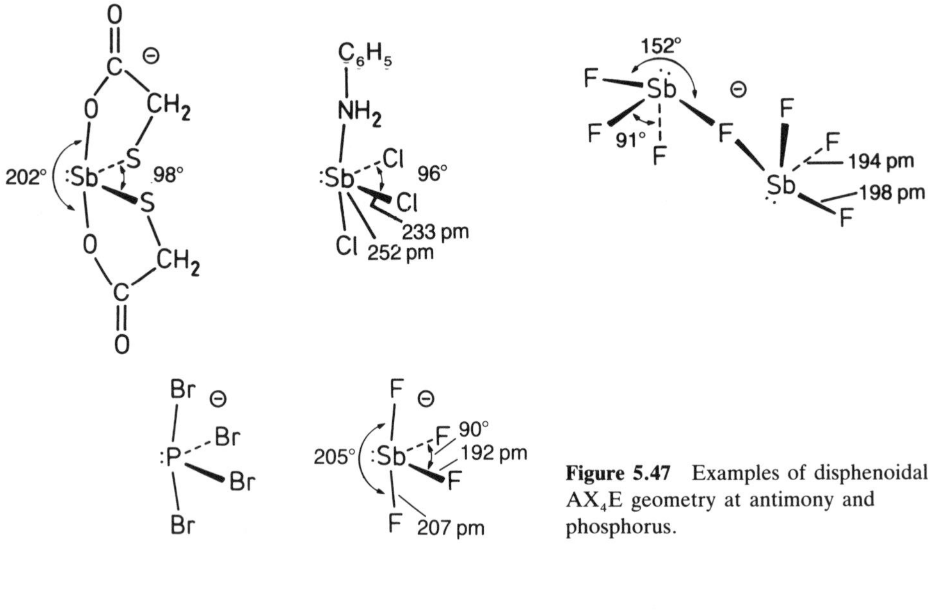

Figure 5.47 Examples of disphenoidal AX_4E geometry at antimony and phosphorus.

the bridging fluorine in one of the axial positions (Figure 5.47). In each of the examples in this figure the deviations produced by the lone pair from the ideal bond angles of 90° and 120° can be clearly seen.

AX_5 Trigonal Bipyramidal Geometry

With the exception of $Sb(C_6H_5)_5$, all the known AX_5 molecules of the group 5 elements have the expected trigonal bipyramidal shape. In all cases where the bond lengths have been measured, the axial bonds are longer than the equatorial bonds, and if there are different ligands, in almost every case the more electronegative ligands occupy the axial positions.

The changes in geometry that occur when, for example, the fluorine atoms in PF_5 are successively substituted by other ligands, such as chlorine, methyl, and trifluoromethyl (Table 5.14), can all be readily interpreted on the basis of the

TABLE 5.14 BOND LENGTHS (pm) AND BOND ANGLES (°) IN SUBSTITUTED PHOSPHORUS PENTAHALIDES

Parameters (Fig. 5.48)	PF_5	PF_4Cl	PF_3Cl_2	PF_2Cl_3	$PFCl_4$	PCl_5
PF_{ax}	157.7(5)	158.1(4)	159.3(4)	159.6(2)	159.7(4)	—
PF_{eq}	153.4(4)	153.5(3)	153.8(7)	—	—	—
PCl_{eq}	—	200.0(3)	200.2(3)	200.5(3)	201.1(3)	202.3(3)
PCl_{ax}	—	—	—	—	210.7(6)	212.7(3)
$X_{eq}PX_{eq}$	120	117.8(7)	121.8(4)	120	120.0(1)	120
$X_{ax}PX_{ax}$	90	90.3(4)	90.0(3)	90	90.9(2)	90

Parameters (Fig. 5.49)	PF_5	CH_3PF_4	$(CH_3)_2PF_3$	$(CH_3)_3PF_2$
PF_{ax}	157.7(5)	161.2(4)	164.3(3)	168.5(1)
PF_{eq}	153.4(4)	154.3(4)	155.3(6)	—
PC	—	178.0(5)	179.8(4)	181.3(1)
CPF_{eq}	—	121(1)	118(1)	—
$F_{ax}PF_{ax}$	90	89.1(4)	89.9(3)	—

Parameters (Fig. 5.50)	PF_4CF_3 CF_3(eq)	PF_4CF_3 CF_3(ax)	$PF_3(CF_3)_2$	$PF_2(CF_3)_3$
PF_{ax}	157.4(7)	154.7(7)	—	160.1(4)
PF_{eq}	153.8(5)	153.8(5)	156.0(3)	—
PC_{eq}	188.2(8)	—	—	188.9(4)
PC_{ax}	—	190(2)	188.5(6)	—

Parameters (Fig. 5.51)	PCl_5	PCl_4CF_3	$PCl_3(CF_3)_2$	$PCl_2(CF_3)_3$
PCl_{ax}	212.7(3)	208.4(9)	—	—
PCl_{eq}	202.3(3)	202.3(3)	203.8(2)	205.5(6)
PC_{ax}	—	198(1)	195(1)	194.4(5)
PC_{eq}	—	—	—	194.4(5)
$Cl_{eq}PCl_{eq}$	—	119.8(1)	—	—
$C_{ax}PCl_{eq}$	—	92.4(2)	—	—
$C_{ax}PC_{ax}$	—	—	—	96(2)

VSEPR model. On successive substitution of fluorine in PF_5 by chlorine, the less electronegative chlorines first take equatorial positions and go into the axial positions only when all three equatorial sites have already been occupied (Figure 5.48). This is the case even for the PF_3Cl_2 molecule, which would possess a higher symmetry if the two chlorines were to take the two axial positions rather than two of the three equatorial positions.

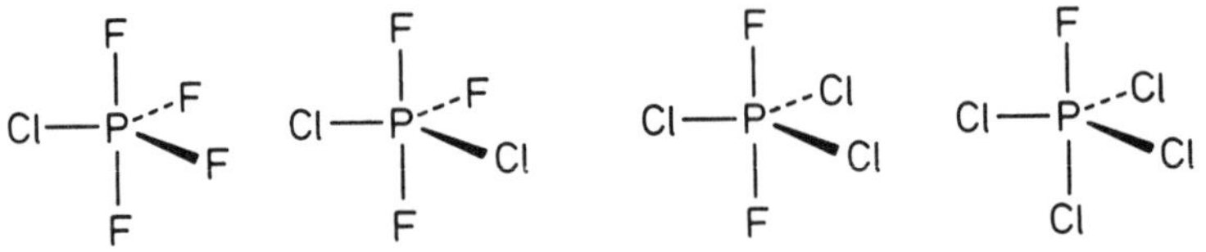

Figure 5.48   Trigonal bipyramidal fluoridechlorides of phosphorus(V) in which chlorine always preferentially occupies the equatorial positions.

The positions taken up by the methyl groups if they are substituted for fluorine in PF_5 again fully correspond to the expectations based on the electronegativity difference between fluorine and the methyl group. Figure 5.49 shows the three known compounds of this type. With increasing number of methyl

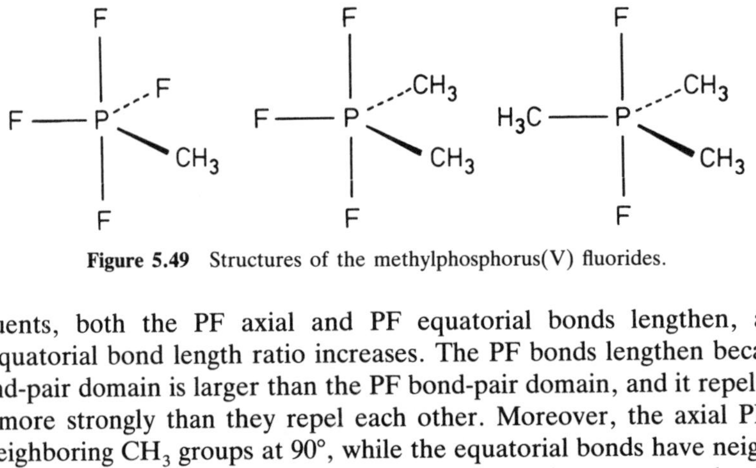

Figure 5.49   Structures of the methylphosphorus(V) fluorides.

substituents, both the PF axial and PF equatorial bonds lengthen, and the axial/equatorial bond length ratio increases. The PF bonds lengthen because the PC bond-pair domain is larger than the PF bond-pair domain, and it repels the PF bonds more strongly than they repel each other. Moreover, the axial PF bonds have neighboring CH_3 groups at 90°, while the equatorial bonds have neighboring CH_3 groups at 120°; so the axial bonds are repelled more strongly than the equatorial bonds, and this causes the axial/equatorial ratio to increase. Also, the PC bonds themselves lengthen gradually. In CH_3PF_4 the FPC bond angles in the equatorial plane are larger than 120°, reflecting the stronger PC bond–PF bond repulsion compared to the PF bond–PF bond repulsions. Upon the second methyl substitution, the PC–PC interaction is stronger than the PC–PF interaction and, accordingly, the CPC bond angle is greater than 120° and greater than the FPC angles.

Substitution of fluorine by the trifluoromethyl group, which is believed to have a slightly smaller electronegativity, should also result in an equatorial position for the CF_3 group. However, the electronegativity difference is very small so the difference in energy between the two isomers will also be very small.

Thus it is not surprising that the experimental results on the structure of PF_4CF_3 are somewhat ambiguous. Some spectroscopic measurements are consistent with one isomer, others are more consistent with the other isomer, while electron diffraction detected a mixture of both isomers. The experimental data may also be interpreted by assuming a rapid interconversion between the two isomers (pseudorotation). In $(CF_3)_3PF_2$ all three trifluoromethyl groups are in equatorial positions, as expected. But in $(CF_3)_2PF_3$ both CF_3 groups were detected at the axial sites, giving the most symmetrical structure (Figure 5.50). Because fluorine

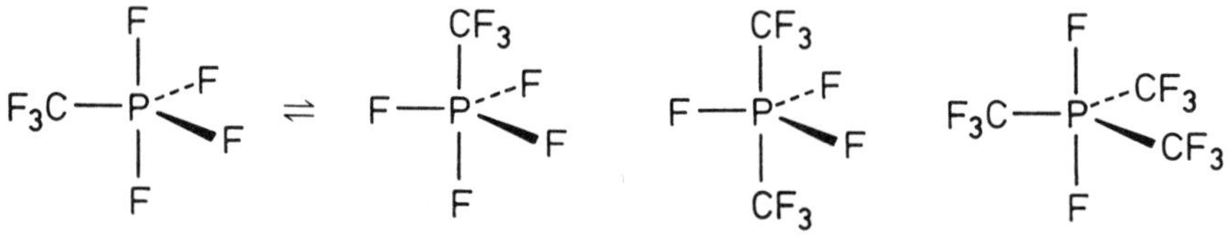

Figure 5.50   Structures of CF_3PF_4, $(CF_3)_2PF_3$, and $(CF_3)_3PF_2$.

and the CF_3 group have very similar electronegativities it is not surprising that in some of these molecules factors other than the electronegativity difference, such as the size of the CF_3 group, may determine the geometry.

If, starting with phosphorus pentachloride, the chlorine ligands are successively substituted by trifluoromethyl groups (Figure 5.51 and Table 5.14), the CF_3 group, which is more electronegative than chlorine, is found to preferentially occupy the axial sites as expected.

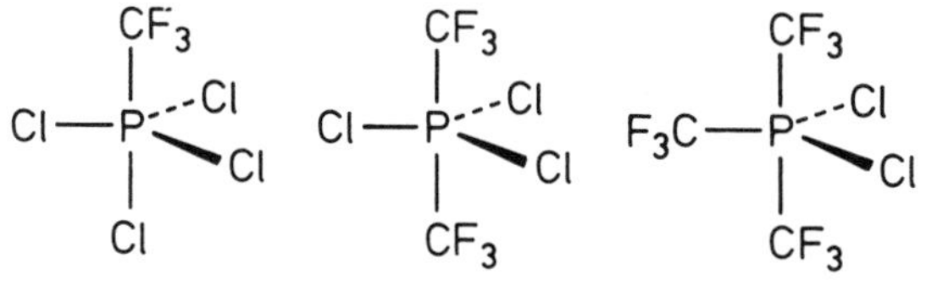

Figure 5.51   Structures of CF_3PCl_4, $(CF_3)_2PCl_3$, and $(CF_3)_3PCl_2$.

An interesting feature of phenyltetrafluorophosphorane (Figure 5.52) is that the benzene ring is almost perpendicular to the equatorial plane. The orientation of the benzene ring can be understood by assuming that there is some delocalization of electrons from the benzene ring to the phosphorus, in other words some double-bond character to the PC bond. Then, using the bent-bond model of the double bond, Figure 5.52 shows that there is an octahedral AX_6 arrangement of bonds around phosphorus, with the benzene ring perpendicular to the equatorial plane. In the amino derivatives of fluorophosphorane the amino group is perpendicular to the equatorial plane for the same reason (Figure 5.52).

Figure 5.52   Bent bond models of the structures of $C_6H_5PF_4$ and F_4PNH_2 showing why the phenyl group and the NH_2 group are perpendicular to the equatorial plane.

When an atom with an AX$_5$ trigonal bipyramidal geometry forms part of a small ring, the most electronegative ligands do not then always preferentially occupy the axial sites. For example, in the dimer of N-methyltrifluorophosphinimine, $(CH_3NPF_3)_2$, which has a four-membered ring with alternating phosphorus and nitrogen atoms (Figure 5.53), the pentacoordinated phosphorus has one axial and two equatorial fluorines and one axial and one equatorial nitrogen. The axial PN bond is 14 pm longer than the equatorial bond. That both nitrogens are not in equatorial positions, as expected, is probably because the small angles in the four-membered ring are more compatible with the 90° N_{ax}—P—N_{eq} bond angle than with the N_{eq}—P—N_{eq} bond angle of 120°.

Other group 5 molecules that have the AX$_5$ trigonal bipyramidal shape include AsF$_5$, SbCl$_5$ (Figure 5.54a), $Sb(C_6H_5)_4OH$, $Sb(C_6H_5)_4OCH_3$, $Sb(C_6H_5)_3(OCH_3)_2$, $Sb(CH_3)_4Cl$, and $Bi(C_6H_5)_4Cl$. But $Sb(C_6H_5)_5$ is the only known example of a group 5 AX$_5$ molecule that has a square pyramid shape (Figure 5.54b), although even in this case it is somewhat distorted toward a trigonal bipyramid. The CSbC angles are 147° and 163° compared to 120° and 180° for a trigonal bypyramid. Presumably because of the small energy difference

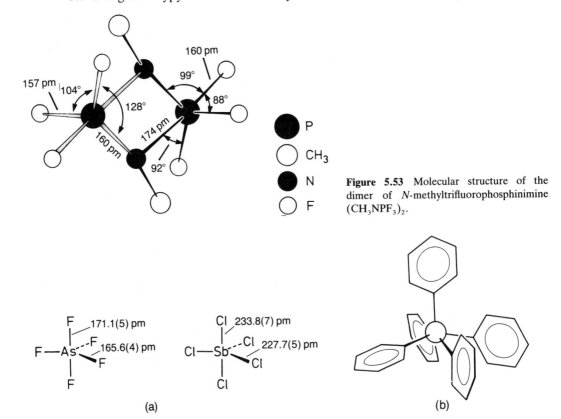

Figure 5.53 Molecular structure of the dimer of N-methyltrifluorophosphinimine $(CH_3NPF_3)_2$.

Figure 5.54 (a) Trigonal bipyramidal AX$_5$ molecules of arsenic and antimony. (b) Antimony pentaphenyl in the solid state is the only example of a square pyramidal AX$_5$ molecule of a group 5 atom.

between these two structures, intermolecular interactions in the solid, or possibly intramolecular interactions between the phenyl groups, are sufficient to favor the approximate square pyramid structure in this particular case.

AX$_5$E Square Pyramidal Geometry

Examples of molecules with this geometry include anions of the general type MX_5^{2-}, such as SbF_5^{2-}, $SbCl_5^{2-}$, $BiCl_5^{2-}$ (Figure 5.55), and some polymeric anions. For example, in $KSbF_4$ the anion is tetrameric, $Sb_4F_{16}^{4-}$, and consists of a ring of four square pyramids joined by linear fluorine bridges (Figure 5.55). The $Bi_2Cl_8^{2-}$

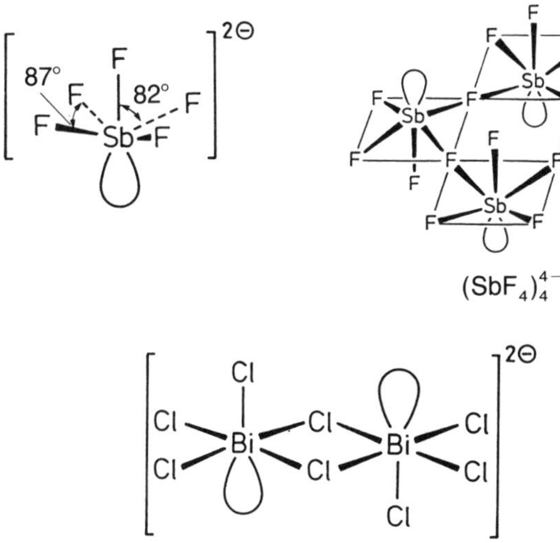

$(SbF_4)_4^{4-}$

Figure 5.55 Examples of molecules in which a group 5 atom has a square pyramidal AX$_5$E geometry.

ion has two bridging chlorines and an AX$_5$E square pyramidal geometry around each Bi (Figure 5.55). In almost all AX$_5$E molecules the four ligands in the base of the square pyramid lie in a plane slightly above the central atom, and the bond angles are all less than 90° because of the larger size of the lone-pair domain than the bond-pair domains, in other words, because the lone-pair–bond-pair repulsions are greater than the repulsions between the bond pairs. As a consequence, the bonds adjacent to the lone pair are longer than the bond opposite the lone pair. There are also several neutral complexes of $SbCl_3$ with the general formula $SbCl_3.2L$, such as $SbCl_3[(C_6H_5)_3AsO]_2$ and $SbCl_3(C_6H_5NH_2)_2$, that have the AX$_5$E geometry.

AX$_6$ Octahedral Geometry

Molecules exhibiting this geometry include many AX_6^- anions and many donor–acceptor complexes in which the group 5 halide is the Lewis acid. Examples include PF_6^-, PCl_6^-, $Sb(OH)_6^-$, and SbF_6^- and many donor–acceptor complexes such as the adduct of PF_5 and pyridine (Figure 5.56). The crystal structure of this adduct shows that the F_aPF angles are all slightly larger than 90° with an average

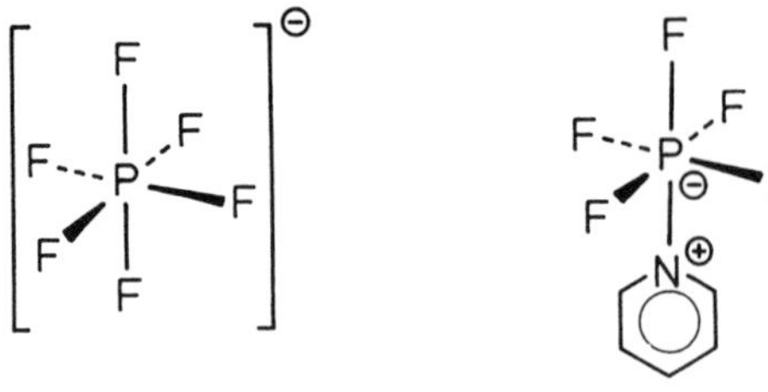

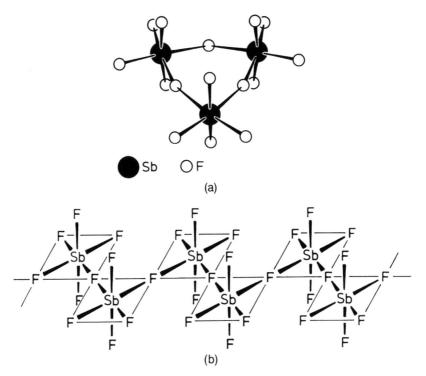

Figure 5.56 Molecules in which phosphorus has an AX$_6$ octahedral geometry.

value of 91.8°. These deviations from the ideal octahedral bond angle are consistent with the P—N bond being longer and weaker than the P—F bonds. The complexes of SbCl$_5$ with POCl$_3$ and PO(CH$_3$)$_3$ have an octahedral geometry that is distorted in the same way as in PF$_5$.pyridine.

Antimony pentafluoride has a trimeric structure in the gas phase in which each SbF$_6$ group has an AX$_6$ octahedral geometry and is *cis* bridged to neighboring SbF$_6$ groups to give a six-membered ring with a chair conformation (Figure 5.57). In the crystalline state there are similar tetramers. The anions Sb$_2$F$_{11}^-$ and Sb$_3$F$_{16}^-$ have structures that are closely related to those of (SbF$_5$)$_3$ and (SbF$_5$)$_4$ and to the polymeric structure of liquid antimony pentafluoride in which each antimony has an octahedral geometry with *cis* bridging to neighboring SbF$_6$ groups (Figure 5.57).

● Sb   ○ F

(a)

(b)

Figure 5.57 Structure of antimony pentafluoride. (a) Gas phase. (b) Liquid state.

AX$_6$E Distorted Octahedral Geometry

As discussed in Chapter 3, it is difficult to make a certain prediction of the expected geometry of an atom with seven electron pairs in its valence shell because there are three arrangements of seven electron pairs that have very similar energies. However, when one of the seven pairs is a lone pair, it seems very probable that the most likely arrangement will be that in which the lone pair has a minimum number of neighboring bond pairs in order to minimize the number of lone-pair, bond-pair interactions. Thus the most probable geometry for an AX$_6$E molecule is a distorted octahedron based on a monocapped octahedral arrangement of seven electron pairs in which the lone pair is in the capping position and has only three neighboring bond pairs. Although this geometry has been observed for XeF$_6$, the ions SbCl$_6^{3-}$, SbBr$_6^{3-}$, BiBr$_6^{3-}$, and BiI$_6^{3-}$ in the solid state all have either a regular octahedral structure (Figure 5.58) or a structure in which not all the bonds are the same length, but in which there is no apparent distortion from the 90° bond angles of the regular octahedral structure. Although there is evidence from Raman spectra that some of these ions are distorted in solution, the exact nature of the distortion is uncertain. The geometry of these ions is further discussed later in conjunction with the isoelectronic ions SeX$_6^{2-}$ and TeX$_6^{2-}$.

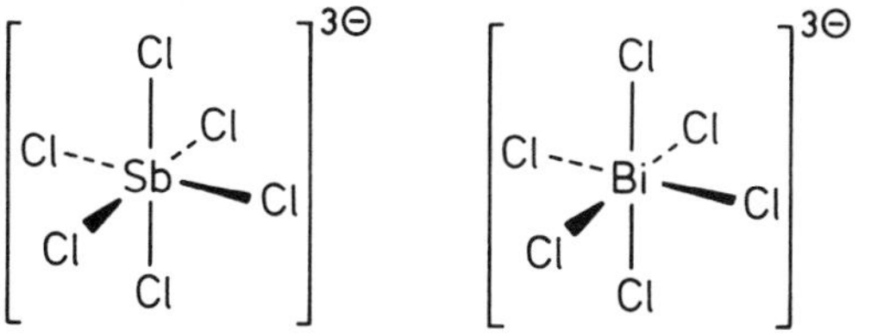

Figure 5.58 Octahedral structures of SbCl$_6^{3-}$ and BiCl$_6^{3-}$.

SULFUR, SELENIUM, AND TELLURIUM

These elements show a great variety of molecular geometries as a consequence of the number of stable oxidation states. The various observed geometries are summarized in Table 5.15.

AX$_2$E Angular Geometry

Sulfur dioxide is a bent molecule with a bond angle of 119.3(1)°, close to the ideal 120°. It appears that the double-bond domains and the lone-pair domain are nearly the same size in this molecule (Figure 5.59). S$_2$O has a similar structure with a slightly smaller bond angle of 118.3(2)°. Other similar molecules include SeO$_2$(g), H$_2$C=S=O, ClN=S=O, CH$_3$N=S=S=NCH$_3$, and (Me$_3$SiN)$_2$S. Thiazyl fluoride and thiazyl chloride also have the angular AX$_2$E geometry with an NS triple bond and a sulfur–halogen single bond (Figure 5.59).

TABLE 5.15 MOLECULAR GEOMETRIES FOR SULFUR, SELENIUM, AND TELLURIUM

$n + m$	Arrangement	n	m	Geometry		Example
3	Trigonal	2	1	AX_2E	Bent	SO_2
		3	0	AX_3	Trigonal	SO_3
4	Tetrahedral	2	2	AX_2E_2	Bent	SCl_2
		3	1	AX_3E	Trigonal pyramidal	$OSCl_2$
		4	0	AX_4	Tetrahedral	O_2SCl_2
5	Trigonal bipyramidal	3	2	AX_3E_2	T-shaped	$[SeC(NH_2)_2]_3^{2+}$
		4	1	AX_4E	Disphenoidal	SF_4
		5	0	AX_5	Trigonal bipyramidal	OSF_4
6	Octahedral	5	1	AX_5E	Square pyramidal	$(CF_3SeCl_3)_2$
		6	0	AX_6	Octahedral	SF_6
7	Monocapped octahedral	6	1	AX_6E	Distorted octahedral[a]	$SeBr_6^{2-}$
	Pentagonal bipyramidal	7	0	AX_7	Pentagonal bipyramidal	$[(C_2H_5)_2NCS_2]_3TeC_6H_5$

n = number of bonds, m = number of lone pairs.
[a]See text.

O=S(119.3°)=O   S=S(118.3°)=O   N≡S(116.9°)=F   **Figure 5.59** Angular AX_2E molecules of sulfur.

AX_3 Trigonal Planar Geometry

Sulfur trioxide, SO_3, has a trigonal planar AX_3 structure with SO double bonds, 141.7(3) pm, that are somewhat shorter than the bonds of sulfur dioxide, 143.1(1) pm. These bond lengths are consistent with the presence of a lone pair on sulfur in SO_2 and the slightly greater effective electronegativity of sulfur in SO_3. Monomeric selenium trioxide, SeO_3, has the same shape (Figure 5.60). $S(NCR_3)_3$, in which there are three S=N double bonds, also has the expected trigonal planar AX_3 geometry.

O=S(=O)(=O)   O=Se(=O)(=O)   **Figure 5.60** Planar triangular AX_3 structures of SO_3 and SeO_3 in the gas phase.

AX_2E_2 Angular Geometry

Bond angles for some simple AX_2E_2 molecules are given in Table 5.16. In general, these are smaller than the ideal tetrahedral angle and are always smaller than the corresponding angle at oxygen. The lone-pair domains tend to occupy as

TABLE 5.16 BOND LENGTHS (pm) AND BOND ANGLES (°) IN AX_2E_2 MOLECULES OF SULFUR, SELENIUM, AND TELLURIUM

Molecule	SX	XSX
SF_2	158.7(1)	98.0(1)
SCl_2	201.5(2)	102.7(2)
SH_2	133.6(3)	92.1(3)
$S(CH_3)_2$	180.7(2)	99.1(1)
$S(PF_2)_2$	213.2(4)	91.3(11)
$S(SiH_3)_2$	212.9(3)	98.4(1)
$S(GeH_3)_2$	220.9(4)	98.9(1)
$S(CF_3)_2$	181.9(30)	97.3(8)

Molecule	SeX	XSeX
$SeCl_2$	215.7(3)	99.6(5)
SeH_2	146.0	90.6
$Se(CH_3)_2$	194.5(1)	96.3(1)
$Se(SiH_3)_2$	227.4(7)	96.6(7)
$Se(GeH_3)_2$	234.4(3)	94.6(5)
$Se(CF_3)_2$	197.8(9)	96(4)

Molecule	TeX	XTeX
$TeCl_2$	232.9(3)	97.0(6)
TeH_2	165.8	90.2
$Te(CH_3)_2$	214.2(5)	94(2)

much space as possible in the valence shell of sulfur, pushing the bond pairs together until their domains begin to overlap and repel each other. Because of the large size of the sulfur valence shell, this overlapping of the bond domains does not occur until the angle between the bond pairs is considerably smaller than in H_2O. That the bond angle is slightly larger than 90° is consistent with the idea that the valence shell of sulfur can accommodate up to six electron pairs in an octahedral arrangement with an inter electron-pair angle of 90°. The slightly smaller bond angles in H_2Se and H_2Te can be attributed to the smaller electronegativity and larger size of these atoms. The bond angles in the series $S(CH_3)_2$, $Se(CH_3)_2$, and $Te(CH_3)_2$ similarly decrease from 99.1° to 96.3° to 94°.

Each sulfur atom in a cyclo S_n molecule has an AX_2E_2 geometry. The bond angle in the crown-shaped S_8 molecule is 108.0° and in S_6, which has a chair conformation, it is 102° (Figure 5.61).

There are some AX_2E_2 molecules containing the N—S—N group that have a bond angle at sulfur that is greater than the tetrahedral angle. Thio-bis(dimethylamine) and N,N'-thio-bis(dicyclohexylamine) are examples with NSN angles of 114(2)° and 110.7(2)°, respectively (Figure 5.61). These relatively large bond angles may be due to the bulkiness of the ligands, but they more probably result from partial delocalization of the nitrogen lone pairs onto the sulfur giving some double-bond character to the bonds.

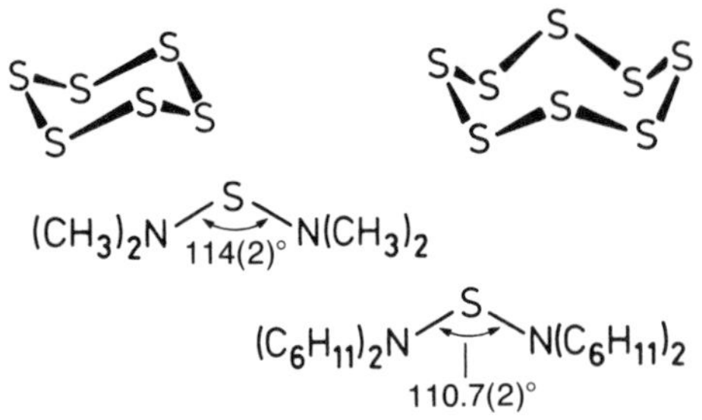

$(CH_3)_2N$ ⟨S⟩ $N(CH_3)_2$
114(2)°

$(C_6H_{11})_2N$ ⟨S⟩ $N(C_6H_{11})_2$
110.7(2)°

Figure 5.61 Angular AX_2E_2 geometry at sulfur in S_6, S_8, $S[N(CH_3)_2]_2$, and $S[N(C_6H_{11})_2]_2$.

AX₃E Trigonal Pyramidal Geometry

Thionyl fluoride, SOF_2 (Figure 5.62), and analogous SOX_2 and $SeOX_2$ molecules are typical examples of the AX_3E geometry. Some structural data are given in Table 5.17; XSX and XSeX angles vary in a narrow range even with bulky

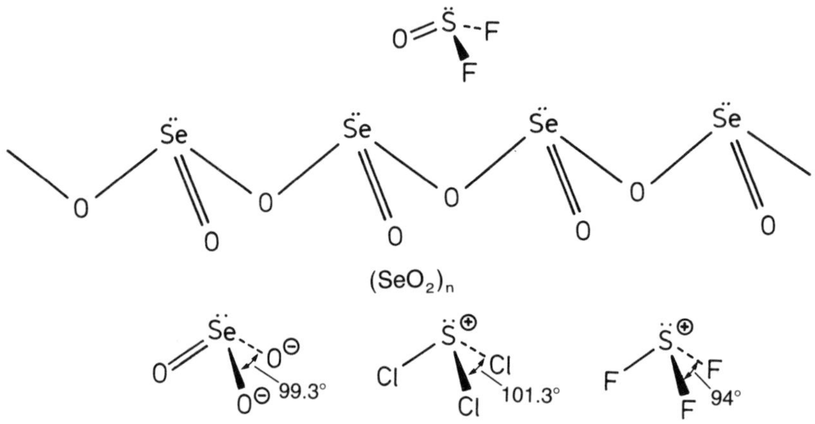

$(SeO_2)_n$

99.3°   101.3°   94°

Figure 5.62 Examples of pyramidal AX_3E geometry in some molecules of sulfur and selenium.

TABLE 5.17   BOND ANGLES (°) IN SULFOXIDE AND SELENOXIDE MOLECULES

Molecule	XSX	OSX
OSF_2	92.3(3)	106.2(2)
$OSCl_2$	96.2(7)	106.4(6)
$OSBr_2$	98.2(2)	107.6(2)
$OS(CH_3)_2$	96.6(2)	106.6(4)
$OS(CF_3)_2$	94.2(8)	104.5(11)
$OS(C_6H_5)_2$	97.3	102.2
$OS[(CH_3)_2N]_2$	96.9(12)	105.5(8)

Molecule	XSeX	OSeX
$OSeF_2$	92.2(1)	104.8(1)
$OSeCl_2$	96.9(7)	106.0(7)

ligands, and the XSX (XSeX) angle is always smaller than the XSO (XSeO) angle. There is a pyramidal arrangement of three oxygen atoms around each selenium in the polymeric structure of crystalline selenium dioxide (Figure 5.62).

The sulfite, selenite, and tellurite ions all belong to the class of AX_3E pyramidal molecules, as do various SX_3^+ cations and the analogous selenium and tellurium ions (Figure 5.62).

The pyramidal AX_3E geometry is also found in several molecules involving Al—S bonding; two examples are shown in Figure 5.63.

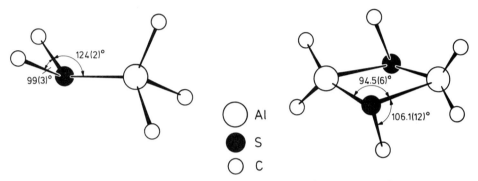

Figure 5.63 Structures of $(CH_3)_2SAl(CH_3)_3$ and $[(CH_3)_2AlSCH_3]_2$.

AX_4 Tetrahedral Geometry

Sulfuryl fluoride, SO_2F_2, and analogous compounds have tetrahedral structures. In all SO_2X_2 molecules, there are three different kinds of bond angles, and any two of them determine the third by virtue of the molecular symmetry. All these angles are listed for several molecules in Table 5.18.

If the two ligands X are different, the molecule is less symmetrical, but the relationship between the three different kinds of bond angles for all known X_2SO_2 and XSO_2Y molecules is always the same:

$$XSY < XSO \text{ or } YSO < OSO$$

TABLE 5.18 BOND ANGLES (°) IN SULFONYL AND SELENINYL MOLECULES

Molecule	XSX	OSX	OSO
O_2SF_2	96.1(2)	108.3	124.0(2)
O_2SCl_2	100.3(2)	108.0(1)	123.5(2)
$O_2S(CH_3)_2$	102.6(9)	108.3	119.7(11)
$O_2S(CF_3)_2$	102.2(8)	107.5	123(3)
$O_2S(OH)_2$	101.3(10)	107.5	123.3(10)
$O_2S(CH_3O)_2$	98.2(8)	108.4(4)	122.3(8)
Molecule	XSeX	OSeX	OSeO
O_2SeF_2	94.1(5)	108.0	126.2(5)

For example, for CH_3SO_2F, $\angle CSF = 98°$, $\angle CSO = 110°$, $\angle OSF = 106°$, and $\angle OSO = 123°$.

The variation of the geometry of the SO_2 group has been examined in a large series of XSO_2Y molecules. With increasing ligand electronegativity, the S=O bond shortens and the OSO angle opens. This is a consequence of the weaker repulsions by the bond pairs of more electronegative ligands. It is interesting that for the molecules shown in Figure 5.64 the O . . . O nonbonded distance stays constant at about 248 pm, while the SO bond lengths and OSO

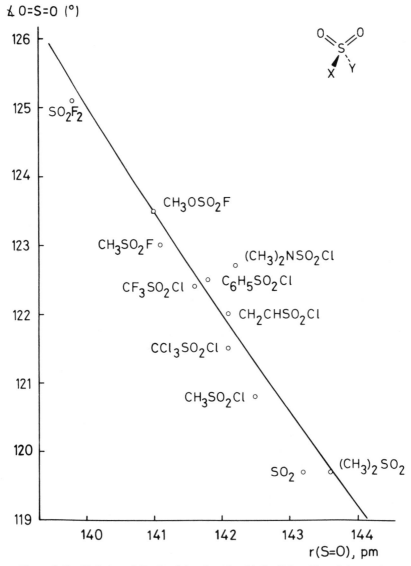

Figure 5.64   Variation of O=S=O bond angle with S=O bond length in a series of $XYSO_2$ molecules.

angles change considerably. Thus it might be argued that it is the nonbonded oxygen–oxygen interactions rather than the bond electron-pair repulsions that determine the geometry of the SO_2 group. However, the 248-pm distance is by no means the shortest nonbonded distance between two oxygen atoms attached to a sulfur atom. It is characteristic only for the O=S=O bonding situation. Thus, for example, dimethyl sulfate, $(CH_3O)_2SO_2$, has three different kinds of OSO bond angles (Figure 3.15) within the same molecule, and the OO nonbonded distance decreases with decreasing SO bond order:

$$O{=}S{=}O \qquad {=}O \ldots O{=} \qquad 248 \text{ pm}$$

$$O{=}S{-}O \qquad {=}O \ldots O{-} \qquad 242 \text{ pm}$$

$$O{-}S{-}O \qquad {-}O \ldots O{-} \qquad 236 \text{ pm}$$

Thus, the nonbonded oxygen atoms get closer to each other with decreasing SO bond order, which clearly indicates the importance of bond domain repulsions in these structures.

Another AX_4 molecule is trifluorosulfur nitride (Figure 5.65). The small FSF bond angles, 94.0(3)°, and large NSF angles, 122.4°, are consistent with the electronegativity of fluorine and the large domain of the SN triple bond. The

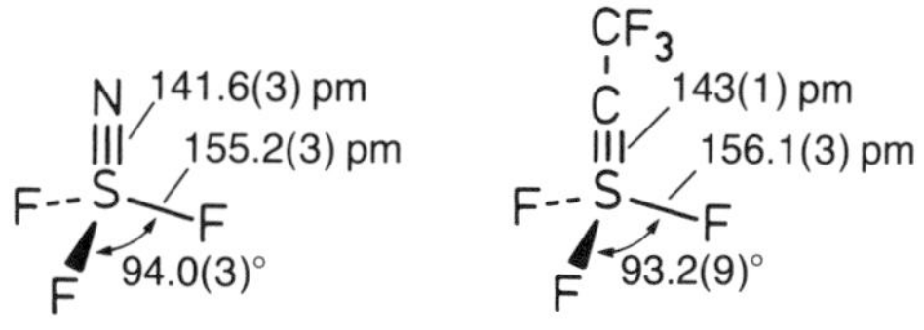

Figure 5.65 Tetrahedral AX_4 molecules, F_3SN and F_3SCCF_3.

geometry of this molecule is analogous to those of OPF_3 and SPF_3, but in each case these molecules have larger FPF angles (Figure 5.45) than the FSF angles because the PO and PS double-bond domains are smaller than the S≡N triple-bond domain. A closely related molecule is trifluoroethylidynesulfur trifluoride, $F_3C{-}C{\equiv}SF_3$ (Figure 5.65). The mean CSF angle is 123° and the mean FSF angle is, accordingly, 93°, which is very similar to the FSF angle in NSF_3. The SC bond is very short, 139.4(5) pm, which is consistent with its formulation as a triple bond.

AX_3E_2 T-Shaped Geometry

The heavy-atom skeleton of the tris(selenourea) ion, $[SeC(NH_2)_2]_3^{2+}$, is shown in Figure 5.66. This T-shaped geometry must result from a trigonal bipyramidal arrangement of five electron pairs around the central selenium, of which two are in the equatorial positions. This must mean that the central selenium has a formal negative charge and that the positive charge is mainly located on the nitrogen atoms, as indicated by the resonance structure in Figure 5.66. This formulation of the bonding is consistent with the observed bond lengths. Thus the SeC bond length corresponds to a normal single bond (193 pm), rather than to a double

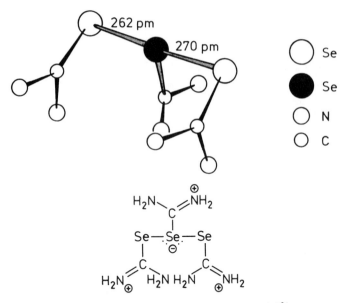

Figure 5.66   Tris(selenourea) ion, $[SeC(NH_2)_2]_3^{2+}$.

bond as in selenourea, and the SeSe bonds are longer than twice the selenium covalent radius (232 pm), as expected for the axial bonds of a trigonal bipyramid. The molecule shown in Figure 5.67 has a similar T-shaped geometry at tellurium.

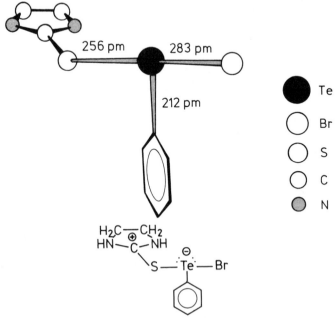

Figure 5.67   Structure of $C_6H_5TeBrSC_3N_2H_6$.

AX$_4$E Disphenoidal Geometry

Sulfur tetrafluoride and selenium tetrafluoride and the related molecules $SF_2(CF_3)_2$ and $SeF_2(CF_3)_2$ have the expected AX$_4$E disphenoidal geometry (Figure 5.68 and Table 5.19). In both cases, as well as in CF_3SF_3, whose

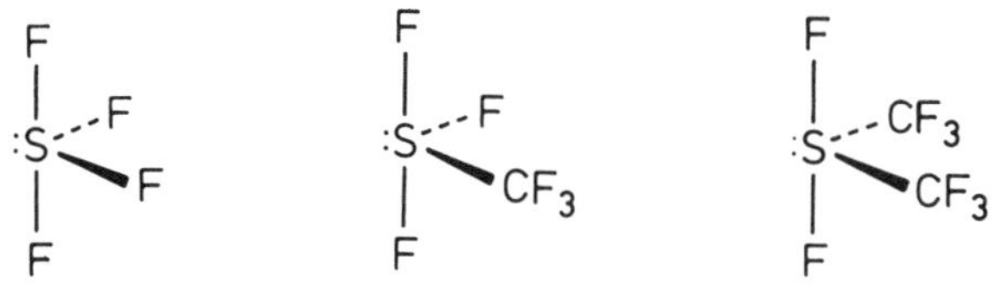

Figure 5.68   Disphenoidal AX$_4$E molecules: SF_4, SF_3CF_3, and $SF_2(CF_3)_2$.

TABLE 5.19   GEOMETRICAL PARAMETERS OF SF_4, SeF_4, $SF_2(CF_3)_2$, and $SeF_2(CF_3)_2$

Parameter	SF$_4$	SF$_2$(CF$_3$)$_2$	SeF$_4$	SeF$_2$(CF$_3$)$_2$
r_{ax} (pm)	164.6(3)	168.1(3)	177(4)	182.7(5)
r_{eq} (pm)	154.5(3)	188.8(4)	168.2(4)	202.2(8)
ax/ax (°)	173.1(5)	173.9(8)	169.2(7)	158(4)
eq/eq (°)	101.6(5)	97.3(8)	100.6(7)	119(2)

geometry has not been determined in detail, the less electronegative CF$_3$ groups are in the equatorial positions. In SF$_4$ and SeF$_4$ the axial bonds are longer than the equatorial bonds, and in all four molecules the ax–ax and eq–eq bond angles are smaller than the ideal values of 180° and 120°, respectively, and except in one case the deviations are greater for the selenium compounds. As we have seen in many other cases, the larger the central atom is the smaller are the bond angles in the presence of a lone pair. However, the CSeC angle in SeF$_2$(CF$_3$)$_2$ is unexpectedly large, and there is no obvious explanation for this anomaly.

In the molecule $CF_3(SF_3)C=C(SF_3)CF_3$ (Figure 5.69), the geometry at sulfur is AX$_4$E disphenoidal, with two axial SF bonds and the third SF bond, the SC bond, and the lone pair in the equatorial positions. The CSF$_{ax}$ bond angles are approximately 86°, the F$_e$SF$_{ax}$ angles are 87° to 88°, and the equatorial CSF angle is 103°. The axial SF bonds are 11 to 13 pm longer than the equatorial SF bond. The $(CF_2SF_2)_2$ dimer has a similar geometry at each sulfur atom (Figure 5.69).

In S$_2$F$_4$ one sulfur has an AX$_4$E disphenoidal geometry and the other an AX$_2$E$_2$ angular geometry, so its formula is best written as F$_3$SSF (Figure 5.70 and Table 5.20). The disphenoidal geometry is considerably distorted but is consistent with the VSEPR model. There is a considerable difference between the two axial bond lengths, and we note that, as expected, the longer axial bond forms smaller angles with the equatorial bond than does the shorter axial bond.

There are many molecules of tellurium in which it has an AX$_4$E disphenoidal geometry, such as dimethyltellurium diiodide, tetraphenyl tellurium (Figure 5.71), biphenyltellurium tribromide, and dibenzotellurophene diiodide (Figure 5.72), in which the halogens are in the axial positions. In biphenyltellurium

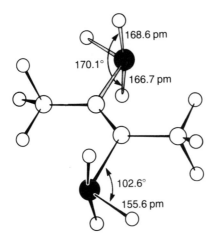

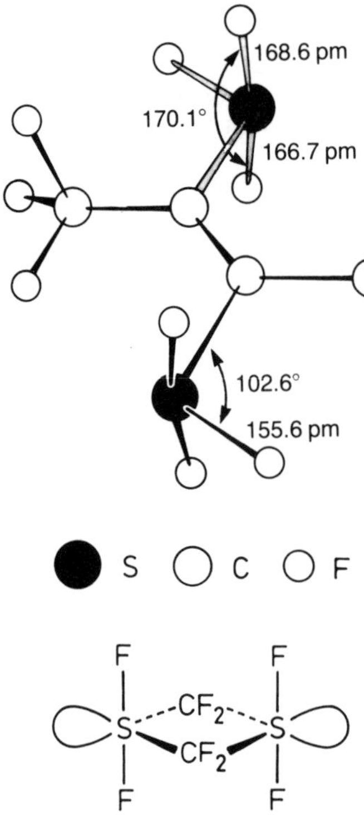

● S   ◯ C   ◯ F

Figure 5.69 Structures of $CF_3(SF_3)C{=}C(SF_3)CF_3$, and of $(SF_2CF_2)_2$.

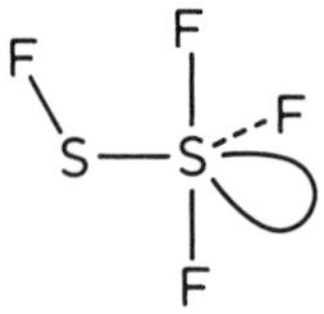

Figure 5.70 Structure of S_2F_4.

TABLE 5.20 GEOMETRICAL PARAMETERS OF THE $FSSF_3$ MOLECULE

Bond Lengths (pm)		Bond Angles (°)	
SF_{ax}	162.5(6)	$F_{ax}SS$	92.2(6)
SF'_{ax}	172.2(8)	$F'_{ax}SS$	76(1)
SF_{eq}	156.9(8)	$F_{eq}SS$	105(1)
		$F_{ax}SF_{eq}$	90(1)
		$F'_{ax}SF_{eq}$	84(3)

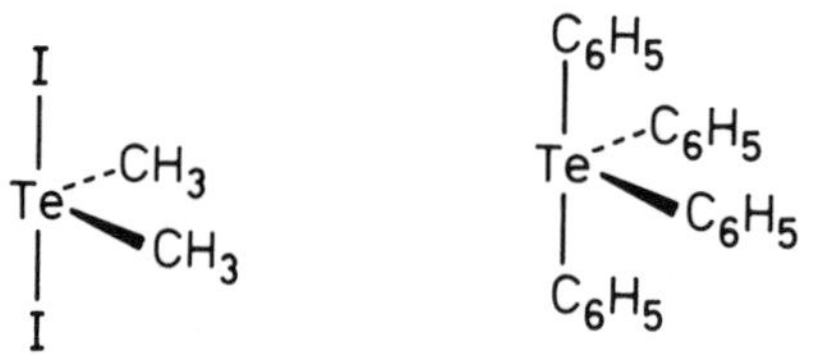

Figure 5.71 Structures of $TeI_2(CH_3)_2$ and $Te(C_6H_5)_4$.

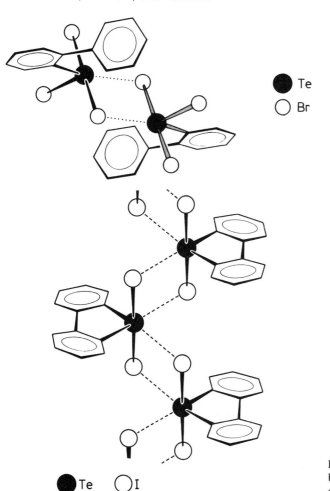

Te
Br

Te ◯ I

Figure 5.72 Structures of biphenyltellurium tribromide and dibenzotellurophene diiodide.

tribromide the molecules are associated into dimers by secondary bonds between a bromine of one molecule and a tellurium of the other, giving an approximate AX_4YE square pyramidal geometry around tellurium. In dibenzotellurophene diiodide the molecules are associated into chains by secondary bonds, giving an approximately octahedral AX_4Y_2E geometry around each tellurium.

AX_5 Trigonal Bipyramidal Geometry

Several compounds of sulfur have molecules with this geometry (Figure 5.73 and Table 5.21). In each case the axial bonds are longer than the equatorial bonds, and the doubly bonded oxygen, nitrogen, and carbon atoms occupy an equatorial position, their large domains causing the expected deviations from the ideal bond angles. It is also of interest to consider these molecules with a doubly bonded ligand as being based on the octahedral arrangement of six electron pairs around sulfur, two pairs being used to form the double bond (Figure 5.73). On the basis

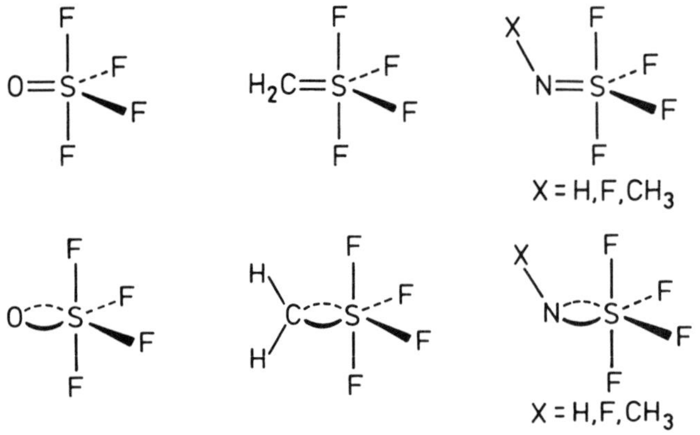

X = H,F,CH₃

X = H,F,CH₃

Fig. 5.73 Molecules of sulfur with a trigonal bipyramidal AX_5 geometry.

TABLE 5.21 GEOMETRICAL PARAMETERS OF OSF_4, H_2CSF_4, AND $OSF_2(CF_3)_2$ MOLECULES

Parameter	OSF_4	H_2CSF_4	$OSF_2(CF_3)_2$
SF_{ax} (pm)	159.5(3)	159.5(15)	164.1(4)
SF_{eq} (pm)	153.8(3)	157.5(15)	—
SC_{eq} (pm)	—	155(2)	189.1(5)
S=O (pm)	140.8(4)	—	142.2(7)
$F_{ax}SF_{ax}$ (°)	164.6(1)	170(2)	173.1(6)
$F_{eq}SF_{eq}$ (°)	112.8(4)	97(2)	97.8(2)

of this model we do not expect a difference in length between the axial and equatorial S—F bonds, and in fact these bonds differ in length less than in SF_4. Alternatively, we may say that, although a lone-pair domain and a double-bond domain in the valence shell of a sulfur atom are approximately the same size, the double-bond domain is less symmetrical and occupies a larger area in the equatorial plane than in the axial plane, so it increases the length of an axial bond less than an equatorial lone pair. The unsymmetrical nature of the S=C double bond also explains why the $F_{eq}SF_{eq}$ angle in $H_2C=SF_4$ is considerably smaller than in OSF_4. The bent-bond model also provides a simple explanation for the observation that the CH_2 and NX groups are perpendicular to the equatorial plane of the trigonal bipyramid (Figure 5.73).

AX_5E Square Pyramidal Geometry

The sulfur pentafluoride ion, SF_5^-, and its Se and Te analogs have the AX_5E square pyramidal geometry (Figure 5.74). In TeF_5^- the bonds in the base of the

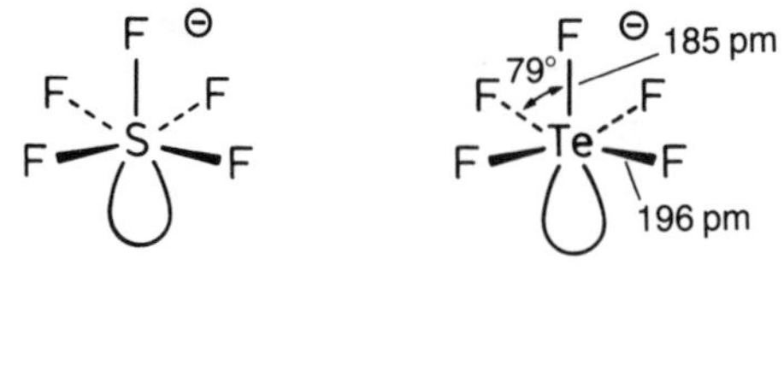

Figure 5.74 AX_5E square pyramidal SF_5^- and TeF_5^- ions.

square pyramid are longer than the axial bond and the bond angles are smaller than 90°. Selenium also has the AX_5E geometry in the dimer of trifluoromethylselenium trichloride, CF_3SeCl_3 (Figure 5.75).

Crystalline tellurium tetrafluoride has a chain structure in which TeF_5 groups are linked by bridging fluorines in such a way that alternate pyramids are oriented in opposite directions. The bonds in the base of the square pyramid are bent away from the lone pair so that the Te atom is located some 30 pm below the plane of the four fluorine atoms and all the bond angles are smaller than 90° (Figure 5.76).

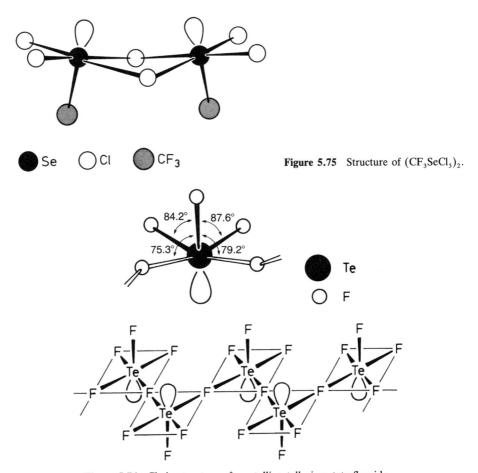

● Se ○ Cl ● CF_3

Figure 5.75 Structure of $(CF_3SeCl_3)_2$.

● Te ○ F

Figure 5.76 Chain structure of crystalline tellurium tetrafluoride.

AX_6 Octahedral Geometry

SF_6, SeF_6, and TeF_6 have the expected regular octahedral geometry (Figure 5.77). Substituting one of the fluorines in SF_6 by another ligand distorts the octahedron as, for example, in sulfur pentafluoridechloride, SF_5Cl (Figure 5.77).

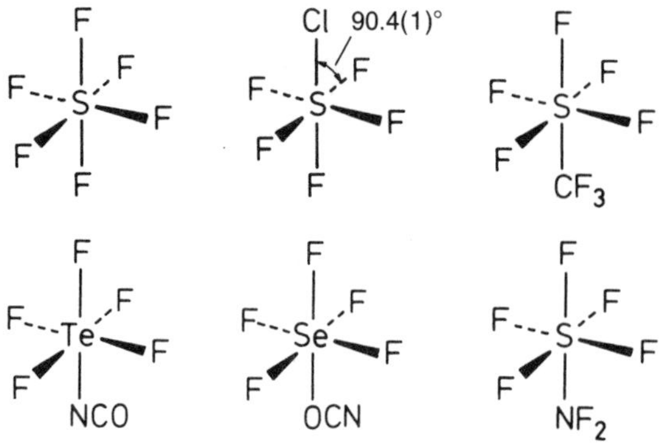

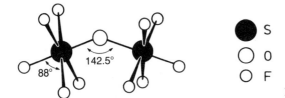

Figure 5.77 Octahedral molecules of sulfur, selenium, and tellurium.

In this molecule the ClSF angles are slightly larger than 90° as a consequence of the smaller electronegativity of chlorine and the larger repulsion exerted therefore by the Cl—S bond pair. The lengths of the two kinds of SF bonds, however, cannot be distinguished on the basis of the available experimental data. In the series of molecules SF_5OSF_5 (Figure 5.78), SeF_5OSeF_5, and TeF_5OTeF_5 and in SF_5CF_3, SF_5NF_2, S_2F_{10}, and SF_5NCO and their selenium and tellurium analogs (Figure 5.77), the four equatorial bonds in each case are bent away from the substituting atom, as expected.

● S
○ O
○ F

Figure 5.78 Structure of F_5SOSF_5.

Telluric acid, $Te(OH)_6$, has an octahedral structure, and in the corresponding oxide TeO_3 there are octahedral TeO_6 groups sharing corners to form a three-dimensional lattice.

AX₆E Distorted Octahedral Geometry

In a previous section we noted that the AX_6E molecules of antimony and bismuth, such as $SbBr_6^{3-}$, do not have the expected distorted octahedral geometry but have a regular octahedral geometry or one in which the bond lengths but not the angles are distorted. The same is true for the isoelectronic ions of selenium and tellurium, such as $SeBr_6^{2-}$ and $TeCl_6^{2-}$. Thus in these ions there appears to be no lone pair in the valence shell. It seems reasonable to assume that with increasing size of the central atom the tendency for the lone pair to be drawn in toward, and to spread around, the core is enhanced to such an extent that in these molecules it is drawn inside the valence shell, surrounding the core in an s-type orbital and effectively becoming the outer shell of the core. This tendency is also enhanced by the presence of six bonding pairs in the valence shell, which leave rather little space for the lone pair. With the addition of the nonbonding electron pair the size of the core increases, and the core charge decreases to +4 from +6 with the result that the bond pairs move farther from the central nucleus, thus increasing the bond lengths. It is a notable feature of these ions that the bonds are considerably longer than expected from the sum of the covalent radii (Table 5.22).

TABLE 5.22 BOND LENGTHS IN SOME AX_6E^3 AND AX_6E^{2-} IONS

	Bond Length (pm)	
Ion	Observed	Calculated (sum of covalent radii)
$SbBr_6^{3-}$	279	255
$SbCl_6^{3-}$	264	240
$TeBr_6^{2-}$	262	249
$TeCl_6^{2-}$	254	234
$SeCl_6^{2-}$	241	216
$SeBr_6^{2-}$	254	231

AX_7 Pentagonal Bipyramidal Geometry

Tellurium is seven coordinated in crystalline tris(diethyldithiocarbamato)phenyltellurium(IV) with six bonds to sulfur atoms and the seventh to a phenyl group (Figure 5.79). The geometry is distorted pentagonal bipyramidal with the TeC bond in one of the axial positions.

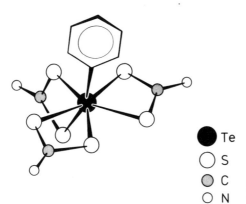

- Te
- S
- C
- N

Figure 5.79 Tris(diethyldithiocarbamato)phenyltellurium(IV).

CHLORINE, BROMINE, AND IODINE

A large number of different molecular geometries are possible for these elements because of the large number of oxidation states that they display. The known molecular geometries are summarized in Table 5.23.

TABLE 5.23 MOLECULAR GEOMETRIES FOR CHLORINE, BROMINE, AND IODINE

$n + m$	Arrangement	n	m	Geometry		Example
3	Trigonal	2	1	AX_2E	Bent	ClO_2^+
4	Tetrahedral	2	2	AX_2E_2	Bent	ClF_2^+, ClO_2^-
		3	1	AX_3E	Trigonal pyramidal	$FClO_2$, F_2ClO^+, ClO_3^-
		4	0	AX_4	Tetrahedral	ClO_3F
5	Trigonal bipyramidal	2	3	AX_2E_3	Linear	ClF_2^-
		3	2	AX_3E_2	T-shaped	ClF_3
		4	1	AX_4E	Disphenoidal	ClF_4^+, $ClF_2O_2^-$, ClF_3O
		5	0	AX_5	Trigonal bipyramidal	ClF_3O_2
6	Octahedral	4	2	AX_4E_2	Square planar	ICl_4^-
		5	1	AX_5E	Square pyramidal	ClF_5, ClF_4O^-
		6	0	AX_6	Octahedral	IOF_5
7	Monocapped octahedral	6	1	AX_6E	Distorted octahedral[a]	IF_6^-
	Pentagonal bipyramidal	7	0	AX_7	Pentagonal bipyramidal	IF_7

n = number of bonds, m = number of lone pairs.
[a]See text.

AX_2E Angular Geometry

The cation ClO_2^+ has this geometry with a bond angle of 122° and a bond length of 131 pm (Figure 5.80). This ion is isoelectronic with SO_2 and has a slightly shorter bond and a slightly larger bond angle, as expected from the greater electronegativity of chlorine.

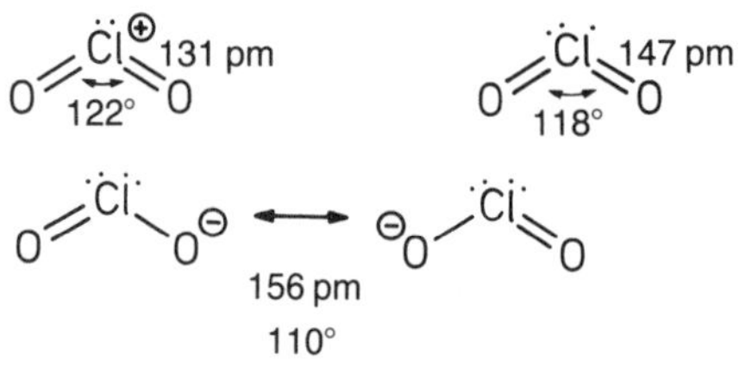

Figure 5.80 Structures of ClO_2^+, ClO_2, and ClO_2^-.

AX_2E_2 Angular Geometry

The halogen cations Cl_3^+, Br_3^+, and I_3^+ are all known to be angular, and the bond length and angle have been determined for I_3^+ (Table 5.24). The analogous interhalogen cations, such as ClF_2^+, are also angular. In each case the bond angle is smaller than the ideal tetrahedral angle (Table 5.24).

The ClO_2^- ion also has this geometry, with a bond angle of 110.5° and a bond length of 156 pm. The bond angle is quite close to the ideal tetrahedral angle that would be expected if the bonds, which presumably have a bond order

TABLE 5.24 BOND LENGTHS AND BOND ANGLES IN THE TRIATOMIC HALOGEN AND INTERHALOGEN CATIONS

Ion	Angle (°)	Bond Length (pm)
I_3^+	101.8	266
BrF_2^+	93.5	169
ICl_2^+	91.5	231

of approximately 1.5, and the lone pairs were to have domains of the same size. When one electron is removed, the stable $ClO_2 \cdot$ free radical is formed. This molecule can be considered to have an AX_2Ee geometry with a lone pair (E) and a single lone electron (e), and so the bond angle, 117.6(1), is larger than in ClO_2^-, but smaller than in ClO_2^+, and the bonds are shorter, 147.3(1) pm, than in ClO_2^-, but longer than in ClO_2^+ (Figure 5.80).

AX_3E Trigonal Pyramidal Geometry

Chloryl fluoride, $FClO_2$, and iodic acid, $IO_2(OH)$ (Figure 5.81), have this geometry. In both cases the largest bond angle is that between the double bonds. The chlorate, bromate, and iodate ions also have this trigonal pyramidal geometry.

AX_4 Tetrahedral Geometry

The perchlorate, perbromate, and periodate ions all have a regular tetrahedral geometry (Figure 5.82). The decrease in the bond length in the series ClO_2^- (156 pm), ClO_3^- (146 pm), and ClO_4^- (142 pm) is consistent with the increasing bond order in this series and the increased effective electronegativity of chlorine with the increased number of attached oxygen atoms. The ClO_3F and $ClO_3(OH)$ molecules have approximately tetrahedral structures in which the largest bond angles are the angles between the Cl=O double bonds (Figure 5.83). In the oxide Cl_2O_7 both chlorine atoms have an approximate tetrahedral geometry quite similar to that of the perchloric acid molecule. The cation $F_2ClO_2^+$ is also tetrahedral, but the detailed molecular parameters are not known.

Figure 5.81 Chloryl fluoride, ClO_2F, and iodic acid, HIO_3.

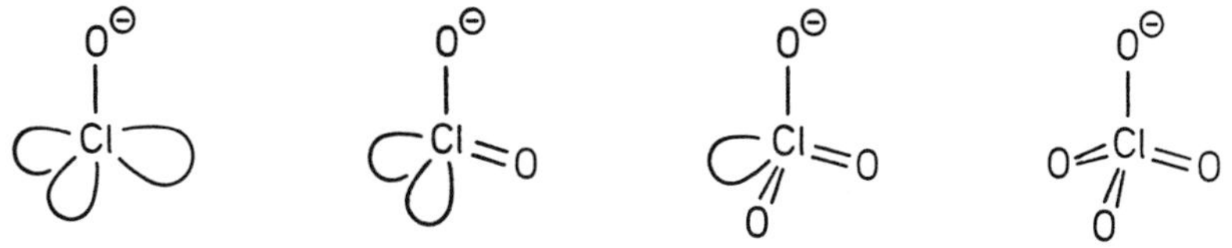

Figure 5.82 Oxoanions of chlorine.

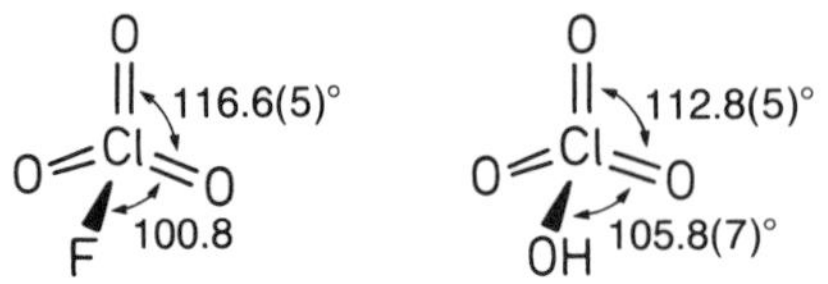

Figure 5.83 Structures of perchloric acid and perchloryl fluoride.

AX_2E_3 Linear Geometry

The triatomic anions of the halogens, such as ICl_2^-, $IBrCl^-$, I_3^-, and Br_3^-, have this geometry (Figure 5.84). There are also numerous solid complexes of iodine and bromine with the compositions $Hal_2.L$ and $Hal_2.L_2$, where L is one of a large variety of donor ligands, which have the linear AX_2E_3 geometry at the halogen atom.

Figure 5.84 The ion ICl_2^- has a linear AX_2E_3 geometry.

AX_3E_2 T-Shaped Geometry

The interhalogen compounds ClF_3, BrF_3, and phenyliodinedichloride, $C_6H_5ICl_2$, have this shape based on a trigonal bipyramidal arrangement of five electron pairs with two lone pairs in equatorial positions. The ClF_3 and BrF_3 molecules show the expected deviations from the ideal shapes (Figure 5.85). The ICl_3 molecule is not known as a monomer, but only in the dimeric form in which the iodine atoms have a square planar AX_4E_2 geometry, as discussed below.

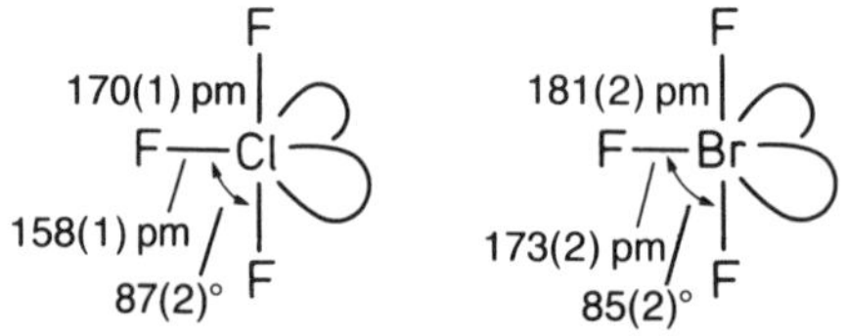

Figure 5.85 T-shaped AX_3E_2 structures of ClF_3 and BrF_3.

AX_4E Disphenoidal Geometry

The ions ClF_4^+, BrF_4^+, and IF_4^+ are isoelectronic with SF_4, and they have the same AX_4E disphenoidal shape (Figure 5.86) in which the axial bonds are longer than the equatorial bonds, and the bond angles are smaller than the ideal values of 90°

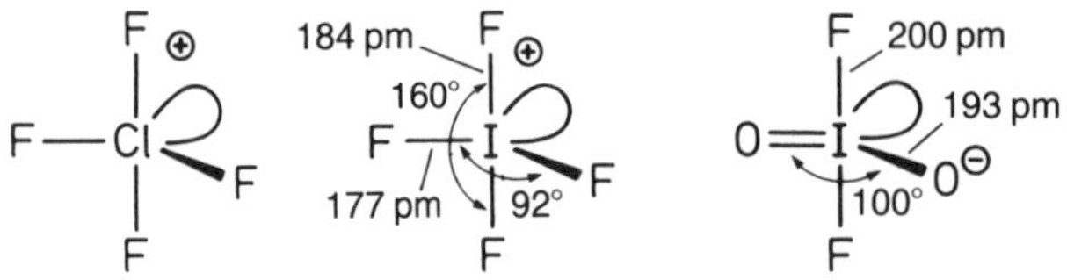

Figure 5.86 Ions with a disphenoidal AX_4E geometry.

and 120°. The oxofluoroanions $IO_2F_2^-$ and $ClO_2F_2^-$ and the oxidefluoride $ClOF_3$ also have this shape (Figure 5.86), although in the latter two cases only the general shape has been established by vibrational spectroscopy. In each case the double-bonded oxygen atoms are in equatorial positions.

AX_5 Trigonal Bipyramidal Geometry

It has been shown by vibrational spectroscopy that chlorine trifluoride dioxide, ClF_3O_2, has the AX_5 trigonal bipyramidal shape with both Cl=O bonds in the equatorial plane (Figure 5.87).

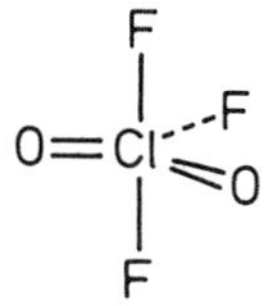

Figure 5.87 Trigonal bipyramidal geometry of ClO_2F_3.

AX_4E_2 Square Planar Geometry

The ICl_4^-, ClF_4^-, BrF_4^-, and IF_4^- ions have this square planar geometry, as do both the iodine atoms in I_2Cl_6 (Figure 5.88). In the latter molecule the bonds to the bridging chlorines are longer, 270 pm, than those to the terminal chlorines, 238 pm, as is found in many other cases. Hence the angle between the terminal bonds (94°) is greater than that between the bridge bonds (84°) at iodine.

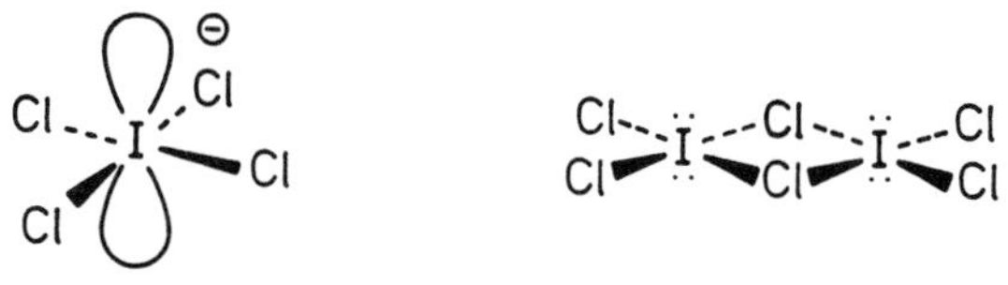

Figure 5.88 Square planar AX_4E_2 geometry of ICl_4^- and I_2Cl_6.

AX_5E Square Pyramidal Geometry

The halogen pentafluorides ClF_5, BrF_5, and IF_5 have this geometry (Figure 5.89). In all cases the central halogen atom is below the basal plane, or, in other words, the basal bonds are tilted away from the lone pair. The lengthening of the basal bonds, compared with the apical bond, is a consequence of lone-pair repulsion. Note also that the bond angle decreases with increasing size of the central atom

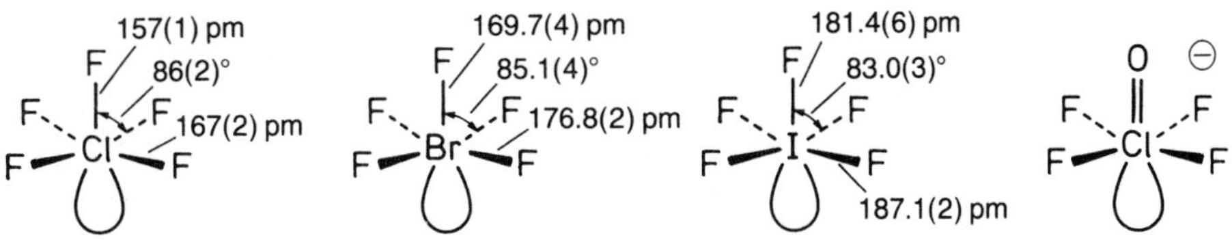

Figure 5.89 Square pyramidal AX_5E halogen pentafluorides and the anion $ClOF_4^-$.

and the consequent decrease in the bond-pair—bond-pair repulsions. The ClF_4O^- ion (Figure 5.89) has a similar square pyramidal structure established from its vibrational spectra. The ClO bond, which is expected to have some double-bond character, is in the apical position.

AX_6 Octahedral Geometry

The IF_6^+, $IO_2F_4^-$, and IO_6^{5-} ions all have an octahedral shape (Figure 5.90). The IF_5O molecule has a distorted octahedral shape because the large domain of the I=O double bond causes the OIF angles to increase to $98.0(3)°$ from the $90°$ of the regular octahedral geometry (Figure 5.90).

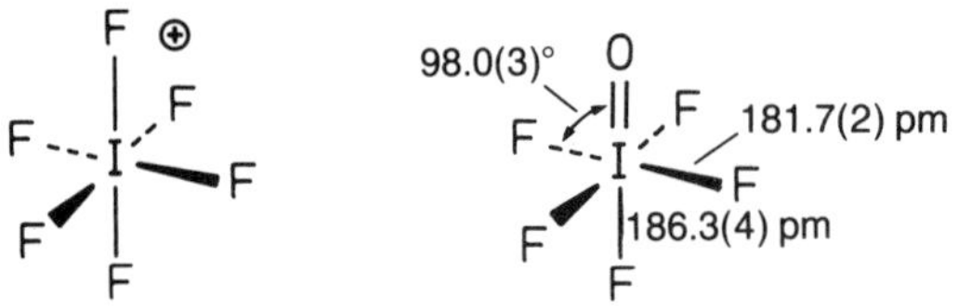

Figure 5.90   Octahedral molecules IF_6^+ and IOF_5.

AX_6E Distorted Octahedral Geometry

The ions IF_6^- and BrF_6^- are of this type, but it is not known with certainty if they have a regular or distorted octahedral geometry. However, in view of the existence of IF_7 it seems probable that there would be sufficient space in the valence shell of iodine to accommodate six bonding and one nonbonding pair, in which case IF_6^- would not be expected to have a regular octahedral structure. The most probable geometry is that which minimizes the number of lone-pair, bond-pair interactions, namely, the $1:3:3$ monocapped octahedral structure with the lone pair in the middle of a distorted octahedral face. For BrF_6^-, vibrational spectroscopic data do not appear to show any appreciable deviation from a regular octahedral structure, so that the lone pair may be in an inner s-type orbital as in $SeCl_6^{2-}$ and analogous ions.

AX_7 Pentagonal Bipyramidal Geometry

Iodine heptafluoride, IF_7, has a pentagonal bipyramidal geometry in the gas phase (Figure 5.91). The axial bonds are shorter, $178.8(7)$ pm, than the equatorial bonds, $185.8(4)$ pm. This difference is a conseqence of the equatorial positions, each of which has two neighbors at $72°$, being more crowded than the axial positions for which the nearest neighbors are at $90°$. In contrast, in the trigonal bipyramid it is the axial positions that are more crowded, and so the axial bonds are longer than the equatorial bonds. The structure of IF_7 is probably, in fact, somewhat distorted from an ideal pentagonal bipyramid geometry because the electron diffraction data can be best interpreted on the basis that there is some slight puckering of the pentagonal plane and a slight decrease in the axial–axial angle from the ideal value of $180°$. Moreover, the molecule is believed to be fluxional and to be converting rapidly from one distorted structure to another.

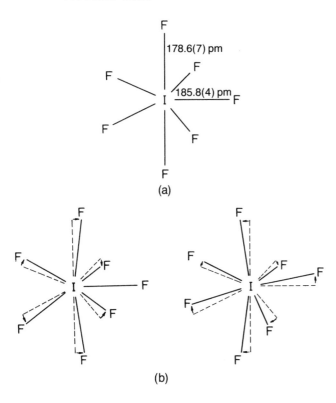

Figure 5.91 (a) IF_7 has an approximately pentagonal bipyramidal geometry. (b) The molecule is probably fluxional with some puckering of the pentagonal plane and a small decrease in the axial–axial angle from the ideal value of 180°.

THE NOBLE GASES

When the noble-gas compounds were first discovered in 1962, it was assumed by many chemists that the bonding in these unexpected compounds must be rather unusual. However, this appears not to be the case. They can all be considered to have normal, if somewhat weak, covalent bonds, and their structures are completely in accord with the principles discussed in this book. Indeed, in several cases the structures of their molecules were correctly predicted on the basis of the VSEPR model before they had been determined experimentally. The molecular geometries of the noble-gas compounds are summarized in Table 5.25.

AX_3E Trigonal Pyramidal Geometry

Xenon trioxide, which is isoelectronic with the iodate ion, IO_3^-, has the same pyramidal AX_3E geometry (Figure 5.92). The $Xe=O$ bonds are short, 176(3) pm, compared with a calculated single-bond length of 196 pm and are best represented as double bonds. The bond angle of 103(2)° is quite small in view of the double-bond character of the XeO bonds. This small bond angle is consistent with the expectation that the effect of a lone pair in decreasing bond angles increases with increasing size of the central atom and the consequent decrease in the strength of the bond-pair—bond-pair repulsions.

TABLE 5.25 MOLECULAR GEOMETRIES FOR NOBLE GASES

$n + m$	Arrangement	n	m		Geometry	Example
4	Tetrahedral	2	2	AX_2E_2	Bent	XeO_2 (predicted)
		3	1	AX_3E	Trigonal pyramidal	XeO_3
		4	0	AX_4	Tetrahedral	XeO_4
5	Trigonal bipyramidal	2	3	AX_2E_3	Linear	XeF_2
		3	2	AX_3E_2	T-shaped	XeF_2O (predicted)
		4	1	AX_4E	Disphenoidal	XeF_2O_2
		5	0	AX_5	Trigonal bipyramidal	XeF_2O_3
		4	2	AX_4E_2	Square planar	XeF_4
6	Octahedral	5	1	AX_5E	Square pyramidal	XeF_4O, XeF_5^+, $[XeFO_3^-]_n$
		6	0	AX_6	Octahedral	XeF_4O_2 (predicted)
7	Monocapped octahedral	6	1	AX_6E	Distorted octahedral	XeF_6, XeF_5O^-
	Pentagonal bipyramidal	5	2	AX_5E_2	Pentagonal planar	XeF_5^-

n = number of bonds, m = number of lone pairs.

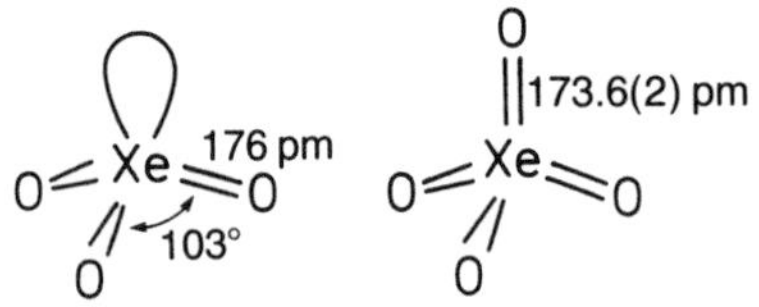

Figure 5.92 Structures of XeO_3 and XeO_4.

AX_4 Tetrahedral Geometry

Xenon tetraoxide has a tetrahedral geometry in the gas phase with a bond length of 174 pm, again indicating that the bonds are best represented as double bonds (Figure 5.92).

AX_2E_3 Linear Geometry

The geometries of krypton difluoride, KrF_2, and xenon difluoride, XeF_2, have been determined in both the gas phase and the crystalline state. Both are linear with bond lengths of 187.5(2) and 197.7(2) pm, respectively, in the gas phase and 188.9 and 200 pm in the solid state (Figure 5.93). The same linear geometry is

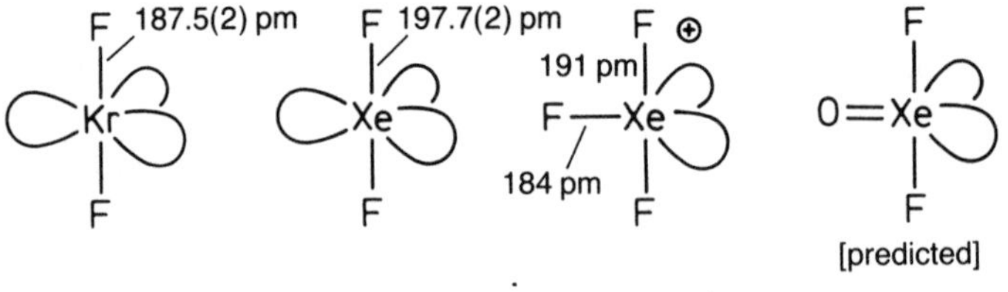

Figure 5.93 Structures of KrF_2, XeF_2, XeF_3^+, and $XeOF_2$.

found in xenon fluoride fluorosulfate $FXeOSO_2F$ (Figure 5.94) and in the compound $FXeN(SO_2F)_2$. The F—Xe—O bond angle in $FXeOSO_2F$ is very slightly smaller than the expected angle of 180°. This deviation from linearity is probably a consequence of intermolecular interactions in the crystal or possibly weak intramolecular interactions.

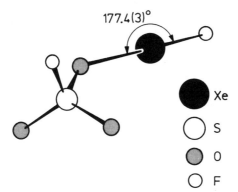

Xe
S
O
F

Figure 5.94 Structure of xenon fluoride fluorosulfate.

AX₃E₂ T-shaped Geometry

The XeF_3^+ ion in the compound $XeF_3.SbF_6$ is an analog of the molecule ClF_3 and has the same T-shaped geometry in which the two axial bonds are longer than the equatorial bond (Figure 5.93). The molecule XeF_2O is not yet known, but it can be predicted to have a T-shaped geometry with the lone pairs and the double bond in the equatorial positions (Figure 5.93).

AX₄E Disphenoidal Geometry

The XeF_2O_2 molecule has been found to have this geometry in the crystal with the lone pair and the double bonds in the equatorial positions (Figure 5.95). The O=Xe=O bond angle is quite small and similar to that in XeO_3. Again we see

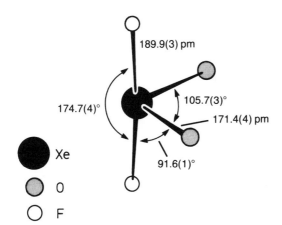

Xe
O
F

Figure 5.95 Structure of XeO_2F_2.

that bond–bond repulsions are weak for a large atom like xenon, and the lone pair is able to push the XeO bonds rather close together even though they are double bonds. An interesting and unusual feature of this molecule is that the axial fluorines are tilted slightly toward the lone pair instead of away from it, as in almost all other molecules of this type. This could be a consequence of the electron density of the double bonds being unsymmetrical and somewhat more concentrated in the direction perpendicular to the equatorial plane than in this plane so that these XeO bonds exert a greater repulsion on the axial bonds than does the lone pair.

AX_4E_2 Square Planar, AX_5E Square Pyramidal, and AX_6 Octahedral Geometries

Xenon tetrafluoride has a square planar shape in both the solid state and in the gas phase (Figure 5.96). The Xe—F bonds have a length of 194 pm.

Xenon oxide tetrafluoride has an AX_5E square pyramidal structure with the short XeO bond in the axial position opposite the lone pair (Figure 5.96). The Xe—F bonds are tilted slightly away from the double bond and toward the lone pair. Perhaps this indicates that in this crowded valence shell containing a total of seven electron pairs the lone pair is beginning to lose its stereochemical activity, as it also does to some extent in XeF_6 (see below).

Potassium monofluoroxenate(VI), $KXeO_3F$, may be considered to contain AX_4E trigonal bipyramidal XeO_3F groups with two of the doubly bonded oxygens in the equatorial positions. However, these units are linked together by fluorine bridges to give a square pyramidal geometry around each xenon, with one double-bonded oxygen in the axial position (Figure 5.97).

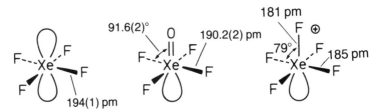

Figure 5.96   Structures of XeF_4, $XeOF_4$, and XeF_5^+.

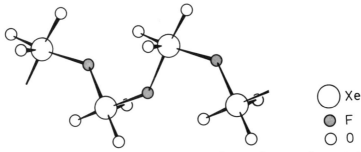

Figure 5.97   Structure of the polymeric XeO_3F^- ion in potassium monofluoroxenate (VI).

XeF$_6$ forms several 1:1 adducts with pentafluorides such as SbF$_5$ that contain the XeF$_5^+$ cation. This has a square pyramidal AX$_5$E geometry in which the bonds in the base are longer than the axial bond and the bond angles are smaller than 90° (Figure 5.96). In the solid state, XeF$_6$ consists of tetramers and hexamers of (XeF$_5^+$)F$^-$ in which the XeF$_5^+$ ion has the same structure as we have just described. These XeF$_5^+$ ions are held together by fluorine bridges formed by the F$^-$ ions (Figure 5.98).

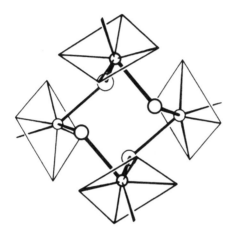

Figure 5.98 Tetramers in the solid-state structure of XeF$_6$.

AX$_6$E Distorted Octahedral Geometry

The molecule XeF$_6$ has a distorted octahedral AX$_6$E geometry with a stereo-chemically active lone electron pair that may be considered to occupy a position in the middle of one face of the octahedron, thus enlarging this face and giving a 1:3:3 or monocapped octahedral structure (Figure 5.99). This structure was predicted on the basis of the VSEPR model some years before direct experimental support for this prediction was obtained. On the basis of molecular orbital arguments, a regular octahedral structure has been predicted. The distortion from a regular octahedral geometry is not large, and it appears that the stereochemical activity of the lone pair is somewhat reduced in this molecule. XeF$_6$ is a nonrigid or fluxional molecule that is continually undergoing a rearrangement that corresponds to the movement of the lone pair between the middle of each face and edge of the octahedron (Figure 5.99). The isoelectronic anion, XeOF$_5^-$, also has a distorted AX$_6$E octahedral structure. Another anion with a similar geometry is (XeOF$_4$)$_3$F$^-$ in which the role of the fifth fluorine is played by the bridging fluorine (Figure 5.100).

AX$_5$E$_2$ Pentagonal Planar Geometry

The XeF$_5^-$ ion has a unique pentagonal planar geometry that arises from the pentagonal bipyramidal arrangement of seven electron pairs. The least crowded

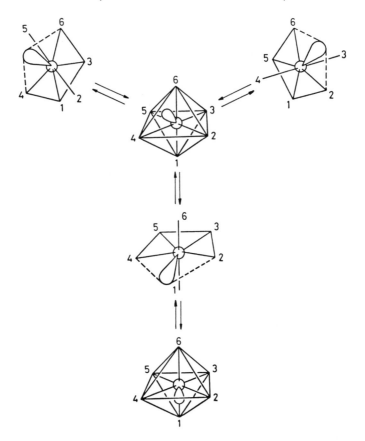

Figure 5.99   XeF$_6$ has a fluxional distorted octahedral AX$_6$E geometry.

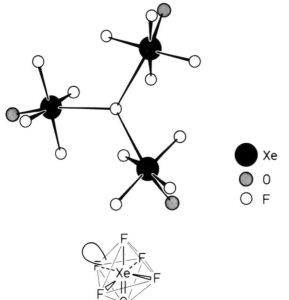

Xe

O

F

Figure 5.100   Structures of (XeOF$_4$)$_3$F$^-$ and XeOF$_5^-$.

positions in this arrangement are the axial positions and so they are occupied by the lone pairs giving a pentagonal planar arrangement for the five bonding pairs.

AX_8E Square Antiprismatic Geometry

The XeF_8^{2-} ion has a regular square antiprismatic geometry despite the presence of the lone pair, which is therefore stereochemically inert. This is perhaps not surprising considering that there are eight bonding pairs of electrons that presumably force the lone pair into the core, as is the case for a number of AX_6E molecules.

REFERENCES AND SUGGESTED READING

F. A. COTTON and G. WILKINSON, *Advanced Inorganic Chemistry*, 5th Ed., Wiley, New York, 1988.

N. N. GREENWOOD and A. EARNSHAW, *Chemistry of the Elements*, Pergamon Press, Oxford, England, 1984.

I. HARGITTAI, *The Structure of Volatile Sulphur Compounds*, Reidel, Dordrecht, 1985.

I. HARGITTAI and M. HARGITTAI, Eds., *Stereochemical Applications of Gas-Phase Electron Diffraction, Part B. Structural Information for Selected Classes of Compounds*, VCH, New York, 1988.

LANDOLT-BÖRNSTEIN, New Series, Volumes II/7 and II/15, *Structure Data of Free Polyatomic Molecules*, Springer, Berlin, 1976 and 1987.

A. F. WELLS, *Structural Inorganic Chemistry*, 5th Ed., Oxford University Press, Oxford, England, 1984.

_6_

The Transition Metals

The transition metals have one or two ns electrons and one or more $(n-1)$ d electrons outside a noble-gas core (Table 6.1). They use the ns electrons and in most cases one or more of their $(n-1)$ d electrons in compound formation. However, the d electrons that are not used for bonding cannot be considered to form part of the valence shell, but rather they constitute a subshell between the valence shell and the core. Although the d electrons are not part of the valence shell, they nevertheless may affect the geometry of a transition-metal molecule.

THE EFFECT OF d ELECTRONS ON GEOMETRY

The valence shell of a transition metal contains only the bonding electron pairs, so to predict the geometry of a molecule of a transition-metal compound we need to consider only the AX_n electron pair arrangements and not the AX_nE_m arrangements. The basic AX_n geometry may, however, be distorted by interaction with the d subshell. Whether or not any distortion is observed depends on whether the d subshell is spherical or if it has a lower symmetry, which depends on the number of electrons in the d subshell. Moreover, distortion is observed only if there is a sufficiently strong interaction between the valence shell and the d subshell. It is not in general possible to predict whether the strength of the interaction between the d subshell and the valence shell is sufficient to distort the basic AX_n shape, but we expect that the strength of this interaction will increase with the number of electrons in the d subshell so that distortions are more frequently observed in cases where the d shell contains more than five electrons than where it contains less than five electrons. We can, however, always predict the basic AX_n shape and how it will be distorted if the interaction with the d subshell is sufficiently strong.

There are three cases for which it can be predicted that no distortion of the basic AX_n geometry will be observed. These are when there are no remaining nonbonding d electrons, or when there are five d electrons each occupying one of

TABLE 6.1 ELECTRONIC CONFIGURATIONS OF THE TRANSITION METALS

		n = 3			4				5			6
		s	p	d	s	p	d	f	s	p	d	s
Sc	[Ne]	$3s^2$	$3p^6$	$3d^1$	$4s^2$							
Ti	[Ne]	$3s^2$	$3p^6$	$3d^2$	$4s^2$							
V	[Ne]	$3s^2$	$3p^6$	$3d^3$	$4s^2$							
Cr	[Ne]	$3s^2$	$3p^6$	$3d^5$	$4s^1$							
Mn	[Ne]	$3s^2$	$3p^6$	$3d^5$	$4s^2$							
Fe	[Ne]	$3s^2$	$3p^6$	$3d^6$	$4s^2$							
Co	[Ne]	$3s^2$	$3p^6$	$3d^7$	$4s^2$							
Ni	[Ne]	$3s^2$	$3p^6$	$3d^8$	$4s^2$							
Cu	[Ne]	$3s^2$	$3p^6$	$3d^{10}$	$4s^1$							
Zn	[Ne]	$3s^2$	$3p^6$	$3d^{10}$	$4s^2$							
Y	[Ne]	$3s^2$	$3p^6$	$3d^{10}$	$4s^2$	$4p^6$	$4d^1$		$5s^2$			
Zr	[Ne]	$3s^2$	$3p^6$	$3d^{10}$	$4s^2$	$4p^6$	$4d^2$		$5s^2$			
Nb	[Ne]	$3s^2$	$3p^6$	$3d^{10}$	$4s^2$	$4p^6$	$4d^4$		$5s^1$			
Mo	[Ne]	$3s^2$	$3p^6$	$3d^{10}$	$4s^2$	$4p^6$	$4d^5$		$5s^1$			
Tc	[Ne]	$3s^2$	$3p^6$	$3d^{10}$	$4s^2$	$4p^6$	$4d^5$		$5s^2$			
Ru	[Ne]	$3s^2$	$3p^6$	$3d^{10}$	$4s^2$	$4p^6$	$4d^7$		$5s^1$			
Rh	[Ne]	$3s^2$	$3p^6$	$3d^{10}$	$4s^2$	$4p^6$	$4d^8$		$5s^1$			
Pd	[Ne]	$3s^2$	$3p^6$	$3d^{10}$	$4s^2$	$4p^6$	$4d^{10}$					
Ag	[Ne]	$3s^2$	$3p^6$	$3d^{10}$	$4s^2$	$4p^6$	$4d^{10}$		$5s^1$			
Cd	[Ne]	$3s^2$	$3p^6$	$3d^{10}$	$4s^2$	$4p^6$	$4d^{10}$		$5s^2$			
Lu	[Ne]	$3s^2$	$3p^6$	$3d^{10}$	$4s^2$	$4p^6$	$4d^{10}$	$4f^{14}$	$5s^2$	$5p^6$	$5d^1$	$6s^2$
Hf	[Ne]	$3s^2$	$3p^6$	$3d^{10}$	$4s^2$	$4p^6$	$4d^{10}$	$4f^{14}$	$5s^2$	$5p^6$	$5d^2$	$6s^2$
Ta	[Ne]	$3s^2$	$3p^6$	$3d^{10}$	$4s^2$	$4p^6$	$4d^{10}$	$4f^{14}$	$5s^2$	$5p^6$	$5d^3$	$6s^2$
W	[Ne]	$3s^2$	$3p^6$	$3d^{10}$	$4s^2$	$4p^6$	$4d^{10}$	$4f^{14}$	$5s^2$	$5p^6$	$5d^4$	$6s^2$
Re	[Ne]	$3s^2$	$3p^6$	$3d^{10}$	$4s^2$	$4p^6$	$4d^{10}$	$4f^{14}$	$5s^2$	$5p^6$	$5d^5$	$6s^2$
Os	[Ne]	$3s^2$	$3p^6$	$3d^{10}$	$4s^2$	$4p^6$	$4d^{10}$	$4f^{14}$	$5s^2$	$5p^6$	$5d^6$	$6s^2$
Ir	[Ne]	$3s^2$	$3p^6$	$3d^{10}$	$4s^2$	$4p^6$	$4d^{10}$	$4f^{14}$	$5s^2$	$5p^6$	$5d^7$	$6s^2$
Pt	[Ne]	$3s^2$	$3p^6$	$3d^{10}$	$4s^2$	$4p^6$	$4d^{10}$	$4f^{14}$	$5s^2$	$5p^6$	$5d^9$	$6s^1$
Au	[Ne]	$3s^2$	$3p^6$	$3d^{10}$	$4s^2$	$4p^6$	$4d^{10}$	$4f^{14}$	$5s^2$	$5p^6$	$5d^{10}$	$6s^1$
Hg	[Ne]	$3s^2$	$3p^6$	$3d^{10}$	$4s^2$	$4p^6$	$4d^{10}$	$4f^{14}$	$5s^2$	$5p^6$	$5d^{10}$	$6s^2$

the five d orbitals, or when there are ten d electrons and the d shell is therefore complete, in other words, when the d-electron configurations are d^0, d^5 (all five orbitals singly occupied), or d^{10} (all five orbitals doubly occupied). In the first case the valence shell surrounds a spherical noble-gas core, and in the other two cases the valence shell surrounds a spherical d subshell and therefore no distortion of the basic AX_n geometry is expected. However, in all other cases there may be a distortion of the basic AX_n geometry, which arises from the repulsion between a nonspherical d subshell and the bonding electron pairs in the valence shell.

If the d subshell is not spherical, then the simplest assumption that we can make about its shape is that it is ellipsoidal and may be either prolate or oblate. The justification for this simplifying assumption can be seen if we consider, for example, the d^9 configuration. If the electron that is missing from the complete spherical d^{10} subshell is from the $d_{x^2-y^2}$ orbital, then the d subshell has an approximately overall prolate ellipsoidal shape, and if it is missing from the d_{z^2} orbital, then the d shell has an approximately overall oblate ellipsoidal shape (Figure 6.1). Although it is only a rough approximation, to assume that a nonspherical d subshell is ellipsoidal, this assumption enables us to easily and correctly predict the distortion of the basic geometries that may be produced by a nonspherical d subshell.

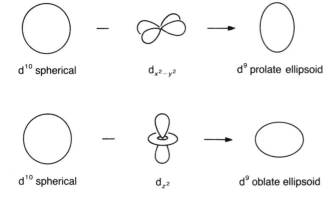

d^{10} spherical $d_{x^2-y^2}$ d^9 prolate ellipsoid

d^{10} spherical d_{z^2} d^9 oblate ellipsoid

Figure 6.1 An incomplete d subshell such as a d^9 subshell can be considered to have an approximately ellipsoidal shape that may be either prolate or oblate.

Thus the shape of a molecule of a transition-metal compound can be predicted by first using the VSEPR model, as we have described for the main-group elements, to obtain the general arrangement of the bonding pairs and therefore the basic geometry of the molecule, and then considering any distortion of this basic shape that might be produced by interaction with the underlying d subshell. We cannot easily predict whether or not any distortion will be observed in a given case, but we can predict what form this distortion will take if the interaction between the valence shell and the d subshell is strong enough.

On the basis of an ellipsoidal shape for a nonspherical d subshell we can make the predictions shown in Table 6.2 and Figure 6.2 for the distortions of the basic molecular shapes that the d subshell may produce.

TABLE 6.2 DISTORTIONS OF THE AX_n GEOMETRIES PRODUCED BY AN ELLIPSOIDAL d SUBSHELL

	Basic Geometry	Distorted Geometry
AX_2	Linear	No distortion
AX_3	Trigonal plane	No distortion
AX_4	Tetrahedron	Disphenoid (elongated or flattened "tetrahedron") or square plane
AX_5	Trigonal bipyramid	Square pyramid (longer axial than basal bonds) or flattened trigonal bipyramid (axial bonds may be shorter than equatorial)
AX_6	Octahedron	Square bipyramid (elongated or flattened "octahedron") or square plane AX_4

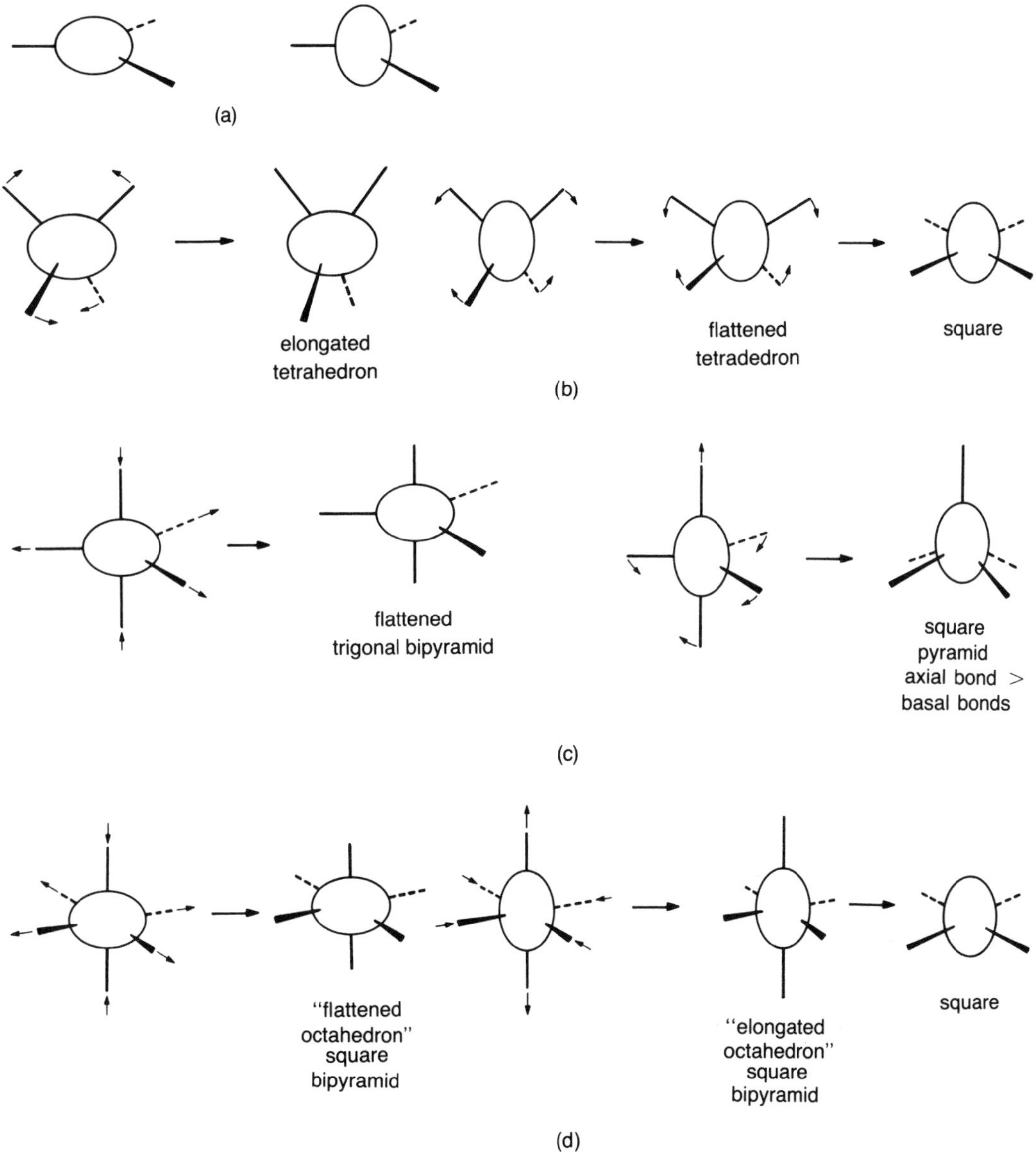

(a)

(b)

elongated
tetrahedron

flattened
tetradedron

square

(c)

flattened
trigonal bipyramid

square
pyramid
axial bond >
basal bonds

(d)

"flattened
octahedron"
square
bipyramid

"elongated
octahedron"
square
bipyramid

square

Figure 6.2 Distortions of the basic AX$_n$ shapes produced by a nonspherical (ellipsoidal) d subshell. (a) An ellipsoidal d subshell has no effect on the trigonal planar AX$_3$ geometry. (b) An oblate ellipsoidal d subshell gives an elongated tetrahedral (disphenoidal) AX$_4$ geometry. A prolate ellipsoidal d subshell gives a flattened tetrahedral (disphenoidal) AX$_4$ geometry or a square planar AX$_4$ geometry. (c) An oblate ellipsoidal d subshell gives a flattened trigonal bipyramidal AX$_5$ geometry. A prolate ellipsoidal d subshell gives a square pyramidal AX$_5$ geometry with the axial bond longer than the basal bonds. (d) An oblate ellipsoidal d subshell gives a flattened octahedral (square bipyramidal) AX$_6$ geometry. A prolate ellipsoidal d subshell gives an elongated octahedral (square bipyramidal) AX$_6$ geometry or a square planar AX$_4$ geometry.

AX$_2$ Linear and AX$_3$ Trigonal Planar Geometries

The repulsion between either a prolate or an oblate ellipsoidal d subshell cannot lead to any distortion of the shape of either a linear AX$_2$ molecule or a trigonal planar AX$_3$ molecule (Figure 6.2).

AX$_4$ Tetrahedral, Disphenoidal, and Square Planar Geometries

Figure 6.2 shows that for a tetrahedral AX$_4$ molecule a prolate ellipsoidal d subshell will repel the bonding electron pairs away from the two ends of the ellipsoid and thus lead to a flattening of the tetrahedron to produce a disphenoidal shape or, in the limiting case, a square planar shape. An oblate ellipsoid would produce an elongation of the tetrahedron. It is not possible to predict which distortion will occur, but the flattening of the tetrahedron to give a square planar shape in the limiting case is observed in the majority of cases. It may be noted that the ligand-field and molecular-orbital treatments of the distortion of the tetrahedral and other basic shapes that treat this distortion in terms of the Jahn–Teller effect similarly cannot predict which type of distortion will occur in any given case.

AX$_5$ Trigonal Bipyramidal and Square Pyramidal Geometries

Figure 6.2 shows that the interaction between a prolate ellipsoidal d shell and the valence-shell bonding electron pairs will cause the bond pairs to avoid the ends of the ellipsoid, and thus the trigonal bipyramid will be destabilized with respect to the square pyramid, which has only one electron pair opposite the end of the ellipsoid. Thus we can predict that when there is a sufficiently strong interaction between the bonding electron pairs and the d subshell the square pyramidal geometry will be preferred to the trigonal bipyramidal geometry. Moreover, the repulsion exerted on the axial electron pair by a prolate ellipsoidal d subshell will cause the axial bond to be longer than bonds in the base of the square pyramid. This difference in bond lengths is in contrast to the situation for AX$_5$E molecules of the main-group elements where, because they interact more strongly with the lone pair, the bonds in the base are longer than the axial bond. The interaction of an oblate ellipsoidal d subshell would further stabilize the trigonal bipyramidal geometry with respect to the square pyramidal geometry, because in the trigonal bipyramid only three bonding pairs are strongly repelled by the d subshell, compared to a strong repulsion of four pairs in the square pyramid. Moreover, the interaction with an oblate d subshell would be expected to reduce the difference in length between the axial and equatorial bonds that is predicted in the case of a main-group element or a transition metal with a spherical d subshell so that this difference might become very small or even negative (equatorial bonds longer than axial bonds). Thus, although we cannot predict which type of distortion will occur in a given case, we can predict that some five-coordinated molecules of a transition element with a nonspherical d subshell will have a square pyramidal shape with longer axial than basal bonds, and that when they have a trigonal bipyramidal shape, the difference between the axial and equatorial bond lengths will be smaller than for a main-group element or a transition metal with a spherical d subshell and may even be zero or negative.

AX_6 Octahedral and Square Bipyramidal Geometries

Figure 6.2 shows that a prolate ellipsoidal d shell will distort an AX_6 octahedral arrangement of bonding electron pairs to give a tall square bipyramid or elongated "octahedral" arrangement, with the two axial pairs at a greater distance from the core than the four equatorial pairs. In the limiting case the two axial ligands may be lost, giving rise to a square planar AX_4 geometry. As we have seen, this shape may also be considered to arise from the distortion of the basic tetrahedral AX_4 shape by a nonspherical d subshell. If the d subshell has an oblate ellipsoidal shape, then the octahedron will be distorted to give a squat square bipyramid or flattened "octahedron," with the two axial bond pairs at a shorter distance from the core than the four equatorial pairs. Again, we cannot predict which type of distortion will occur, but it is found experimentally that the most common distortion is an elongation of the octahedron.

Comparison of the Geometries of Main-Group Element Molecules and Transition Metal Molecules

Application of the VSEPR model to the geometry of transition-metal compounds is basically the same as for a compound of a main-group element. In fact, because there are no lone pairs in the valence shell of a transition metal, only the basic AX_n geometries need to be considered. The deviations from these regular geometries produced by differences in bonding pairs due to differences in ligand electronegativities or to multiple bonding can be discussed in exactly the same way as for a main-group element. The main difference is that the interaction between the valence shell and the d subshell, which may produce a distortion of the basic AX_n geometry, must also be considered. However, it is not possible to predict with certainty whether or not a distortion will be observed in any given case and, if it is, which of two possible forms, such as the elongation or flattening of an AX_6 octahedron, it will take. Thus the VSEPR model has less predictive power for transition-metal compounds than for compounds of the main-group elements, but it is still very useful in that it provides the basis for understanding, in a simple way, many aspects of the geometry of transition-metal compounds. Generally, in the molecular-orbital and ligand-field treatments of the geometry of transition-metal compounds, the basic geometry is assumed and only the distortion of this geometry produced by the d orbitals is considered. Moreover, these theories do not treat deviations from the basic geometry produced by ligands of different electronegativity or by multiple bonding and no theory can predict which of the two possible distortions will occur in any given case.

Finally we should note that for those molecules where there are no nonbonded d electrons and no distortion of the basic AX_n shape is expected, there may nevertheless be sufficient interaction between the bonding electrons and a polarizable core to distort the core and cause small changes in bond angles, just as we discussed in Chapter 5 for the halides and some other compounds of calcium and the heavier alkaline earth metals. These small effects will be discussed in those particular cases where they may have been observed.

TWO COORDINATION

AX_2 Linear Geometry

This geometry is expected for all two-coordinated compounds of the transition metals irrespective of the number of d electrons and the strength of the interaction between the d subshell and the valence shell. Some examples are given in Table 6.3. Linear AX_2 geometry is particularly common for copper, silver, and gold in the +1 oxidation state, for zinc and cadmium in the +2 oxidation state, and for mercury in both the +1 and +2 oxidation states, all of which have d^{10} configurations.

TABLE 6.3 EXAMPLES OF AX_2 LINEAR GEOMETRY

d Subshell	Central Atom	Compounds
d^5	Mn(II)	$MnF_2(g)$, $MnCl_2(g)$, $MnBr_2(g)$
d^6	Fe(II)	$FeCl_2(g)$, $FeBr_2(g)$
d^7	Co(II)	$CoCl_2(g)$, $CoBr_2(g)$
d^8	Ni(II)	$NiF_2(g)$, $NiCl_2(g)$, $NiBr_2(g)$
d^{10}	Cu(I)	Cu_2O, $CuFeO_2$, $CuCrO_2$, $KCuO$, $CuCl$
	Ag(I)	$KAg(CN)_2$, $[Ag(NH_3)_2]_2SO_4$, $AgCN$, $KAgO$, $CsAgO$, Ag_2O
	Au(I)	$KAu(CN)_2$, $AuCl$, AuI, $AuCN$, $CsAuO$, $ClAuPCl_3$, $ClAuP(C_6H_5)_3$
	Zn(II)	$ZnCl_2(g)$, $ZnBr_2(g)$, $ZnI_2(g)$, $Zn(CH_3)_2(g)$, $Zn[Co(CO)_4]_2$, $Zn\{N[Si(CH_3)_3]_2\}_2(g)$
	Cd(II)	$CdBr_2(g)$, $Cd[Co(CN)_4]_2$
	Hg(I)	Hg_2F_2, Hg_2Cl_2, Hg_2Br_2
	Hg(II)	$HgCl_2(g, s)$, $HgBr_2(g, s)$, $HgI_2(g, s)$, HgO, Na_2HgO_2, $HgNH_2Cl$, $HgNH_2Br$, $Hg(CH_3)_2(g, s)$, $Hg(CF_3)_2(g, s)$, $Hg(CH_3S)_2$, $Hg[Co(CO)_4]_2$
	Pd(0)	$Pd\{PC_6H_5[C(CH_3)_3]_2\}_2$

AX_2 linear geometry is observed for only a few Cu(I) compounds, such as crystalline Cu_2O (Figure 6.3), because Cu(I) has higher coordination numbers in most of its compounds. AX_2 linear geometry is more common for Ag(I) and especially for Au(I). The complex cyanide ions $Ag(CN)_2^-$ and $Au(CN)_2^-$, the complex halide ions such as $AgCl_2^-$, and the cation $Ag(NH_3)_2^+$ all have this linear geometry. The cyanides AgCN and AuCN have infinite linear chain structures (Figure 6.4). AuCl, AuI, and AgSCN have zigzag chain structures with two

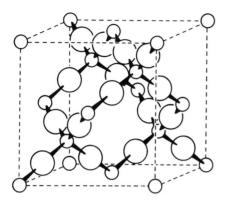

Ag, Cu
O

Figure 6.3 Crystal structure of Cu_2O and Ag_2O.

colinear bonds at each metal atom (Figure 6.4). Silver oxide, Ag_2O, is isostructural with Cu_2O (Figure 6.3).

Zinc and cadmium are two coordinated in only a few compounds in the gas phase, such as the halides, dialkyl compounds, and $Zn[Co(CO)_4]_2$ (Table 6.3 and Figure 6.5). In contrast, mercury is two coordinated in many of its compounds in the solid state as well as in the gas phase (Table 6.3). There are infinite chains with linear mercury coordination in crystalline HgO and in the $(HgNH_2^+)_n$ ion in $HgNH_2Cl$ (Figure 6.6).

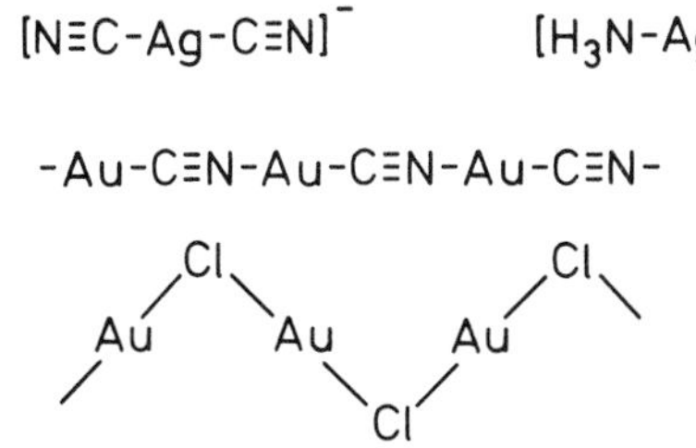

[N≡C-Ag-C≡N]⁻
[H₃N-Ag-NH₃]⁺
-Au-C≡N-Au-C≡N-Au-C≡N-
Cl
Cl
Au
Au
Au
Cl

Figure 6.4 Some ions and polymeric molecules of silver and gold with a linear AX_2 geometry.

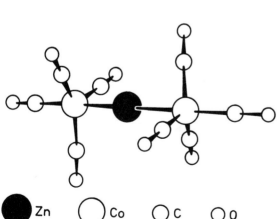

Zn
Co
C
O

Figure 6.5 Structure of $Zn[Co(CO)_4]_2$.

O
O
Hg
Hg
Hg
O
O
NH₂
NH₂
Hg
Hg
Hg
NH₂
NH₂

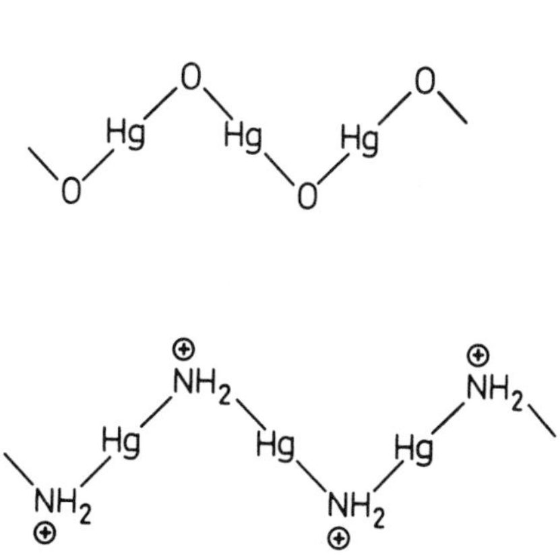

Figure 6.6 Mercury has a linear AX_2 geometry in HgO and $(HgNH_2^+)_n$.

All mercury(I) compounds contain pairs of Hg atoms in a linear
X—Hg—Hg—X arrangement. Figure 6.7 shows the structure of the polymeric
cation in $[Hg_2(OHg)_2](NO_3)_2$ in which infinite —Hg—O—Hg—O— chains are
connected into a layer by Hg—Hg units, and both Hg(I) and Hg(II) have a linear
geometry.

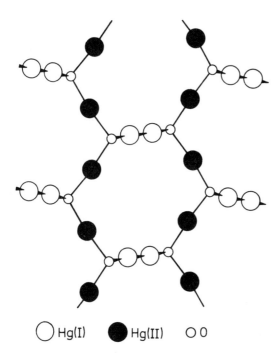

⬭ Hg(I)   ● Hg(II)   ○ O

Figure 6.7 Structure of
$[Hg_2(OHg)_2](NO_3)_2$.

Manganese(II) has a d^5 subshell configuration. Its gas-phase dihalides,
MnF_2, $MnCl_2$, and $MnBr_2$, are linear. All two-coordinated molecules of the
transition metals with nonspherical d subshells are also expected to be linear.
Examples of molecules of this type that have been shown to have a linear
geometry include the gaseous dihalides of Fe(II), Co(II), and Ni(II).

Bis(cyclopentadienyl) and bis(aryl) transition-metal complexes such as ferrocene, $(C_5H_5)_2Fe$, and bis(benzene)chromium, $Cr(C_6H_6)_2$, and related compounds such as $(C_5H_5)CuP(C_6H_5)_3$ have linear AX_2 geometry at the metal
(Figure 6.8). The geometry is independent of the order of the bonds, which in the
case of the bonds to cyclopentadiene and to benzene can be regarded as a type of
triple bond (Figure 4.2).

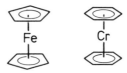

Figure 6.8 Biscyclopentadienyl iron
and bis(benzene)chromium.

THREE COORDINATION

AX₃ Trigonal Planar Geometry

This geometry is expected for all molecules of the transition metals in which the transition-metal atom is three coordinated, irrespective of the number of d electrons or of the strength of the interaction between the d subshell and the valence electrons. Some examples are given in Table 6.4.

TABLE 6.4 EXAMPLES OF AX₃ PLANAR GEOMETRY

d Subshell	Central Atom	Compound
d^1	Ti(III)	$Ti\{N[Si(CH_3)_3]_2\}_3$, $[(C_5ME_5)_2TiN]_2$
d^2	V(III)	$V\{N[Si(CH_3)_3]_2\}_3$
d^3	Cr(III)	$Cr\{N[Si(CH_3)_3]_2\}_3$
d^5	Fe(III)	$FeCl_3(g)$, $Fe\{N[Si(CH_3)_3]_2\}_3$
	Mn(II)	$Mn\{N[Si(CH_3)_3]_2\}_2 \cdot (tetrahydrofuran)$
d^7	Co(II)	$Co\{N[Si(CH_3)_3]_2\}_2 \cdot P(C_6H_5)_3$
d^{10}	Cu(I)	$Cu[P(C_6H_5)_3]_2Br$, $\{Cu[SP(CH_3)_3]_3\}ClO_4$, $Na_2Cu(CN)_3 \cdot 3H_2O$, $C_8H_8 \cdot CuCl$, $C_7H_8 \cdot CuCl$, $[Cu(SP(CH)_3)]Cl\}_3$, $NaCu(CN)_2 \cdot 2H_2O$
	Au(I)	$\{Au[P(C_6H_5)_3]_3\}B(C_6H_5)_4$, $Au[P(C_6H_5)_3]_2Cl$

In the vapor at high temperatures, iron(III) chloride, in which iron has a d^5 configuration, exists as planar monomeric $FeCl_3$ molecules (Figure 6.9). In $Fe\{N[Si(CH_3)_3]_2\}_3$, both iron and nitrogen have a planar geometry (Figure 6.10), but the $N[Si(CH_3)_3]_2$ groups are twisted out of the FeN_3 plane. The planar geometry at nitrogen is presumably due to the delocalization of the nitrogen lone pair into the valence shells of the silicon and iron atoms.

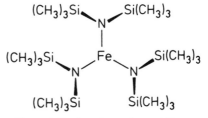

Figure 6.9 Structure of iron(III) chloride in the gas phase at high temperature.

Figure 6.10 Iron has a planar AX₃ geometry in $Fe\{N[Si(CH_3)_3]_2\}_3$.

Three-coordinated molecules and complex ions of Pd(0), Cu(I), Ag(I), Au(I), and Hg(II) in which the metal has a d^{10} configuration all have a planar AX₃ geometry. Examples include $Pd(PPh_3)_3$, $Cu(CN)_3^{2-}$, HgX_3^-, $AgI(PEt_2Ph)_2$,

$AuCl(PPh_3)_2$, $Au(PPh_3)_3^+$ (Figure 6.11), the inifinite chain $[Cu(CN)_2^-]_n$ ion in $KCu(CN)_2$ and $NaCu(CN)_2 \cdot 2H_2O$ (Figure 6.12). The bond angles at copper in these last two compounds are somewhat distorted from the ideal angles of 120°, with the smaller angles being formed by the longer Cu—N bond. In the polymeric cyclooctatetraene complex $(C_8H_8)CuCl$, copper forms three coplanar bonds, two to Cl atoms and one to a double bond of the cyclooctatetraene (Figure 6.13). The bond between Cu and cyclooctatetraene can be regarded as a type of three-center bond such as we described in Chapters 4 and 5 for B_2H_6 and $Al_2(CH_3)_6$ (Figures 4.21 and 5.11). An ellipsoidal d subshell is not expected to distort the AX_3

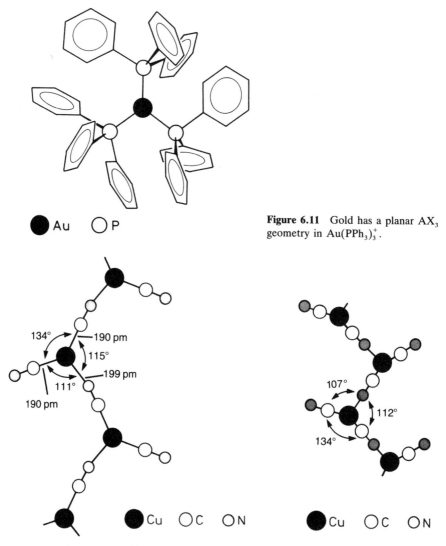

Figure 6.11   Gold has a planar AX_3 geometry in $Au(PPh_3)_3^+$.

Figure 6.12   The infinite chain anion $Cu(CN)_2^-$ in $NaCu(CN)_2 \cdot 2H_2O$ and in $KCu(CN)_2$.

trigonal planar geometry, and no examples of any distortion are known among the compounds of the transition metals. Some dialkylamides of chromium(III) (d^3) have a planar AX_3 geometry. There are also several complexes of V(III) (d^2) and Ti(III) (d^1), including $V[N(SiMe_3)_2]_3$, $V[CH(SiMe_3)_2]_3$, and $Ti[N(SiMe_3)_2]_3$ that have this geometry (see Figure 6.10). Another interesting example is the molecule $(C_5Me_5)_2Ti\text{—}N\equiv N\text{—}Ti(C_5Me_5)_2$ in which titanium has a d^1 subshell and has trigonal planar AX_3 geometry at both titanium atoms (Figure 6.14).

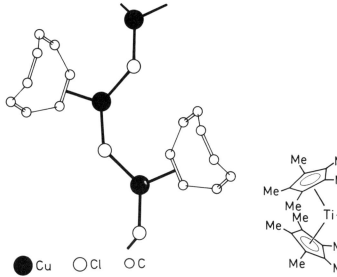

● Cu   ○ Cl   ○ C

Figure 6.13 Polymeric cyclooctatetraene complex, $(C_8H_8)CuCl$.

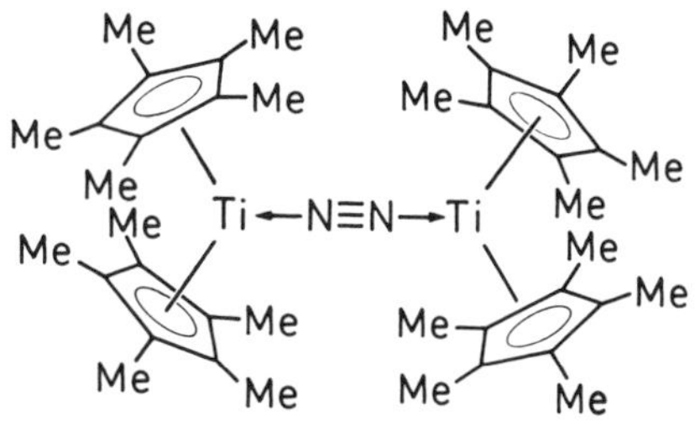

Figure 6.14 Titanium has a trigonal planar geometry in the complex $(C_5Me_5)_2Ti\text{—}N\equiv N\text{—}Ti(C_5Me_5)_2$.

The most probable alternative shape for an AX_3 molecule is the trigonal pyramid, and this shape has been observed for some molecules of the lanthanides, although not for the transition metals. However, because of the involvement of f electrons, discussion of the geometry of molecules of the lanthanides is more complicated than for the transition metals and is beyond the scope of this book.

FOUR COORDINATION

AX_4 Tetrahedral Geometry

Tetrahedral coordination is common for elements of the first transition-metal series. For d^0, d^5, and d^{10} subshells, no distortion of the tetrahedral geometry is expected.

The gaseous tetrahalides of the titanium subgroup are typical tetrahedral molecules. They include TiX_4, ZrX_4, and HfX_4, with X = F, Cl, Br, or I. The bond lengths are collected in Table 6.5.

TABLE 6.5 BOND LENGTHS (pm) IN AX$_4$ TETRAHEDRAL
TETRAHALIDES OF THE TITANIUM SUBGROUP

AX$_4$	F	Cl	Br	I
Ti	175.4(3)	217.0(2)	233.9(5)	254.6(4)
Zr	190.2(4)	232.8(5)	246.5(4)	266.0(10)
Hf	190.9(5)	231.6(5)	245.0(4)	266.2(8)

All the MO_4^{n-} ions of the transition metals with d^0 configurations, such as MnO_4^-, CrO_4^{2-}, VO_4^{3-}, ReO_4^-, and TcO_4^-, have the expected tetrahedral geometry.

The tetraoxides RuO_4 and OsO_4 and the related anion OsO_3N^- are tetrahedral.

Crystalline chromium trioxide, CrO_3, consists of chains of tetrahedra (Figure 6.15). Several polymers of tungsten trioxide have been identified in its vapors, for example, $(WO_3)_2$, $(WO_3)_3$, and $(WO_3)_4$, with ring structures in which the tungsten and oxygen atoms alternate and tungsten has a tetrahedral geometry. Structural details of the trimer are given in Figure 6.16. Chromium trioxide and molybdenum trioxide have similar structures in the vapor phase.

The approximately tetrahedral shape of VOF_3 and $VOCl_3$ (Figure 6.17) is analogous to that of POF_3 and $POCl_3$, but the O=VX angle is smaller than the XVX angle rather than larger than the XVX angle as in the phosphorus analogs, and as predicted by the VSEPR model (Table 6.6). The related molecule $ClN=VCl_3$ has very similar bond angles (Table 6.6). Chromyl fluoride, CrO_2F_2, and chromyl chloride, CrO_2Cl_2 (Figure 6.17), also have an approximately tetrahedral geometry analogous to SO_2F_2 and SO_2Cl_2, but again the OCrO angle is

TABLE 6.6 BOND LENGTHS (pm) AND BOND ANGLES (°) IN SOME
GASEOUS OXOHALIDES OF V(V) AND Cr(VI)

VOX$_3$	VOF$_3$	VOCl$_3$	ClNVCl$_3$	CrO$_2$X$_2$	CrO$_2$F$_2$	CrO$_2$Cl$_2$
V=O (V=N)	157.0(5)	157.0(5)	165.1(6)	Cr=O	157.5(2)	158.1(2)
V—X	172.9(2)	214.2(2)	213.8(2)	Cr—X	172.0(2)	212.6(2)
OVX (NVCl)	107.5(4)	107.6	106.0(6)	OCrO	107.8(8)	108.5(4)
XVX	111.4	111.3(1)	113.4(3)	XCrX	111.9(9)	113.2(3)
				OCrX	109.3(2)	108.7(1)

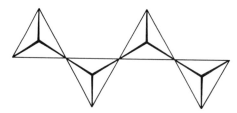

Figure 6.15 Crystalline CrO_3 consists of CrO_4 tetrahedra linked into chains by sharing two corners.

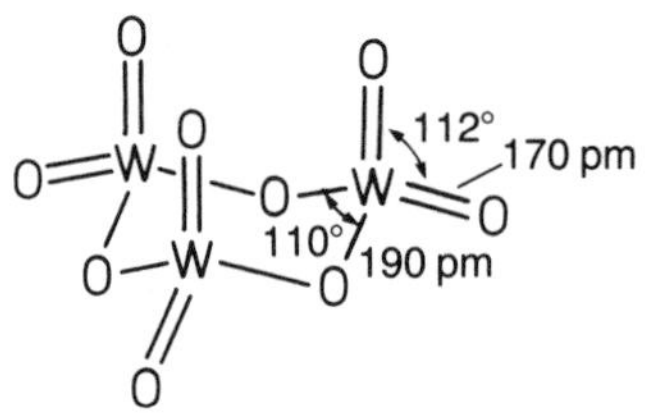

Figure 6.16 Structure of the trimer of tungsten(VI) oxide.

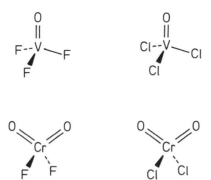

Figure 6.17 Oxohalides of vanadium and chromium.

unexpectedly smaller than the XCrX angle (Table 6.6). However, the calculated electron density distributions that are discussed in Chapter 7 have shown that there is sufficient interaction between the bonding electrons and the core in $VOCl_3$ and in CrO_2Cl_2 to cause a slight distortion of the spherical charge distribution of the core. Each bonding domain produces a small local concentration of charge in the core in the opposite direction from the bond, the largest concentration being produced by the larger double-bond domain. This charge concentration opposite the double bond causes the angle between the V—Cl bonds to increase slightly and to become larger than the OVCl angle. Similar considerations can explain the deviations in the bond angles of CrO_2Cl_2 and CrO_2F_2 (see Chapter 7). The effect that we have described here is essentially the same as we discussed in Chapter 5 to account for the nonlinearity of some of the dihalides of calcium.

Molybdenum and tungsten have a tetrahedral geometry in a number of compounds that contain a metal–metal triple bond, such as $Mo_2[N(CH_3)_2]_6$ (Figure 6.18), $W_2[CH(SiMe_3)_2]_6$, and $Mo_2Cl_2(NMe_2)_4$.

Examples of tetrahedral molecules in which the metal has a d^5 configuration include the ions $FeCl_4^-$ and $MnCl_4^{2-}$, as well as the molecule Fe_2Cl_6, which has a chlorine bridged structure (Figure 6.19) similar to that of dimeric aluminum trichloride (Figure 5.10).

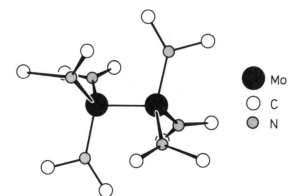

● Mo
○ C
◐ N

Figure 6.18 Molybdenum has a tetrahedral AX_4 geometry in $Mo_2[N(CH_3)_2]_6$ in which there is a metal–metal triple bond.

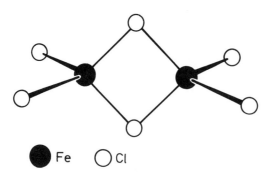

Figure 6.19 Dimeric structure of iron(III) chloride in the gas phase.

In many compounds of Cu(I), Ag(I), Au(I), Zn(II), and Cd(II), which all have a d^{10} configuration, the metal has a tetrahedral AX_4 geometry. Examples include complex ions such as $Cu(CN)_4^{2-}$, $ZnCl_4^{2-}$, $HgCl_4^{2-}$, infinite chain anions such as $(CuCl_3^-)_n$ in $CsCuCl_3$ (Figure 6.20) and $(AgI_3^{2-})_n$ in Cs_2AgI_3, and crystals with the zinc-blende structure, such as ZnS, CuCl, CuBr, CuI, and AgI (Figure 6.21). In $SrZnO_2$, the ZnO_4 tetrahedra form a layer by sharing all vertices (Figure 6.22). The red form of HgI_2 also consists of tetrahedra sharing all corners, but arranged in a different way (Figure 6.22).

Nickel, palladium, and platinum have tetrahedral coordination in their zero oxidation state (d^{10}) compounds, such as $Ni(CO)_4$, $Ni(PF_3)_4$, $Pt(Ph_2PCH_2CH_2PPh_2)_2$, and $Pd(PF_3)_4$ (Figure 6.23). Rhodium and iridium form some tetrahedral complexes, such as $Rh(CO)_4$ and $Ir(CO)_3PPh_3$, in their -1 oxidation state.

Although interaction between a nonspherical d subshell and the valence shell may cause a distortion of the regular tetrahedral geometry, in many molecules of this type no distortion has been observed. Examples include VCl_4 and $V(NEt_2)_4$ (d^1), oxoanions such as CrO_4^{3-} and MnO_4^{2-} (d^1), and complex ions of Co(II) (d^7) and Ni(II) (d^8), such as $CoCl_4^{2-}$, $Co(SCN)_4^{2-}$, and $NiCl_4^{2-}$ (Table 6.7). In other AX_4 molecules, however, the tetrahedral AX_4 geometry is distorted, as described in the following section.

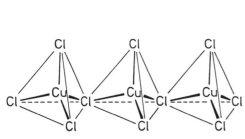

Figure 6.20 Structure of the infinite linear chain anion $(CuCl_3^-)_n$ in $CsCuCl_3$.

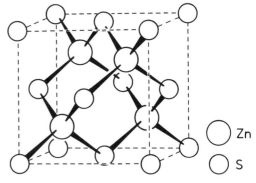

Figure 6.21 Crystal structure of zinc blende (sphalerite).

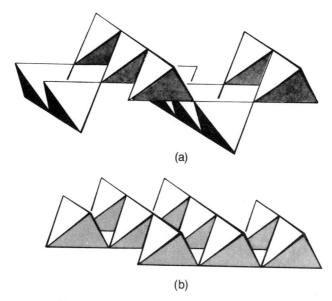

Figure 6.22 (a) In $SrZnO_2$, ZnO_4 tetrahedra are linked by sharing oxygens to give the layer structure shown. (b) The red form of HgI_2 also has a layer structure in which HgI_4 tetrahedra share corners in a different way.

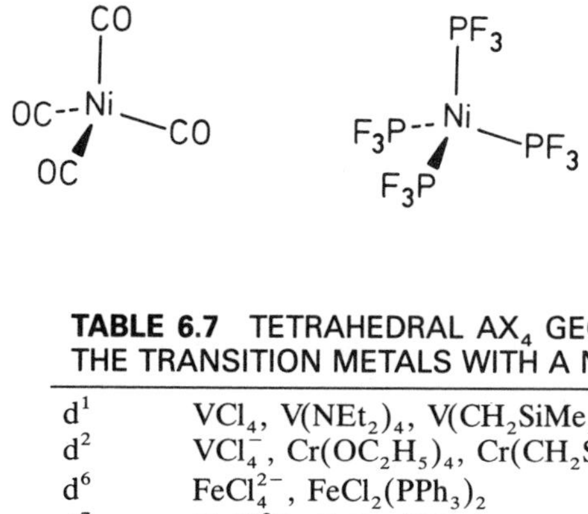

Figure 6.23 Tetrahedral molecules $Ni(CO)_4$ and $Ni(PF_3)_4$.

TABLE 6.7 TETRAHEDRAL AX_4 GEOMETRY IN MOLECULES OF THE TRANSITION METALS WITH A NONSPHERICAL d SUBSHELL

d^1	VCl_4, $V(NEt_2)_4$, $V(CH_2SiMe_3)_4$, CrO_4^{3-}, and MnO_4^{2-}
d^2	VCl_4^-, $Cr(OC_2H_5)_4$, $Cr(CH_2SiMe_3)_4$
d^6	$FeCl_4^{2-}$, $FeCl_2(PPh_3)_2$
d^7	$CoCl_4^{2-}$, $CoBr_2(PPh_3)_2$
d^8	$NiCl_4^{2-}$, $NiCl_2(PPh_3)_2$
d^9	$Ni(PPh_3)_3Br$, $IrNO(PPh_3)_3$

AX₄ Square Planar and Disphenoidal Geometry

Interaction of a prolate ellipsoidal d subshell with a tetrahedral arrangement of four valence-shell electron pairs causes a flattening of the tetrahedron to give a disphenoidal geometry, which in the limiting case becomes square planar (Figure 6.2). This square planar geometry can also be regarded as the limiting case of distortion of an octahedral complex, where the axial ligands are repelled to a very large distance by the d subshell (Figure 6.2).

The disphenoidal or partially flattened "tetrahedral" geometry is rather rare. It is found, for example, in a few Ni(II) (d^8), Cu(II) (d^9), and Cr(II) (d^4) complexes, such as $NiBr_2[P(C_6H_5)_3]_2$, $CrCl_2(MeCN)_2$, $CrI_2(OPPh_3)_2$, and some salts of the $CuCl_4^{2-}$ ion. In the $CuCl_4^{2-}$ ion in the Cs^+ and $(CH_3)_4N^+$ salts, for example, there are four bond angles of about 100° and two of about 125°, instead of six angles of 109.5° (Figure 6.24). But in other salts the $CuCl_4^{2-}$ ion has a square planar geometry or has two other ligands at a greater distance, thus completing a square bipyramidal or elongated "octahedral" geometry.

Square planar AX_4 geometry is most often found in compounds in which the transition metal has a d^8 subshell, as in some Ni(II), Pd(II), Pt(II), Cu(III), and Au(III) compounds, or a d^9 subshell, as in some Cu(II) compounds (Table 6.8).

In crystalline CuO, copper has a square planar AX_4 geometry and oxygen has a tetrahedral geometry (Figure 6.25). Palladium dichloride, $PdCl_2$, forms

Figure 6.24 The $CuCl_4^{2-}$ ion in Cs_2CuCl_4 and $[(CH_3)_4N]_2CuCl_4$ has a flattened "tetrahedral" structure.

TABLE 6.8   EXAMPLES OF AX_4 SQUARE PLANAR GEOMETRIES

d Subshell	Compound
d^8	$K_2Ni(CN)_4$, $NiBr_2[P(C_2H_5)_3]_2$, $PdCl_2$, K_2PdCl_4, $Ca[Pd(CN)_4].5H_2O$, $PtCl_2(NH_3)_2$, $K[PtCl_3NH_3].H_2O$, Au_2Cl_6, AuF_3, $KAuF_4$, $KAu(NO_3)_4$
d^9	$[Ag(pyridine)_4^+]$, CuO, $KCuCl_3$, $Cu(NO_3)_2$, $Cu(acac)_2$, $(NH_4)_2CuCl_4$, $K_2(Cu_2Cl_6)$

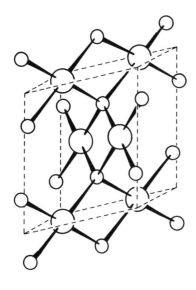

◯ Cu

◯ O   **Figure 6.25** Crystal structure of CuO.

infinite chains in the crystal, with a square planar geometry at palladium (Figure 6.26). $AuCl_3$ exists as dimers in the crystal, with a planar AX_4 geometry around both gold atoms, whereas AuF_3 forms helical chains, retaining, however, the planar geometry around gold (Figure 6.27).

We have seen above that nickel(II) forms both tetrahedral and square planar complexes. In the case of the complex $NiBr_2.[P(CH_2C_6H_5)(C_6H_5)_2]_2$, both tetrahedral and square planar geometries are found in the same crystal structure (Figure 6.28).

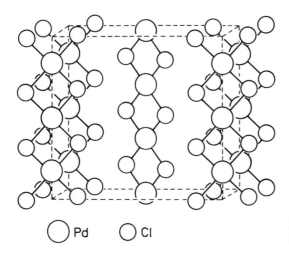

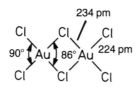

Pd ◯   Cl ◯

Figure 6.26 Crystal structure of $PdCl_2$.

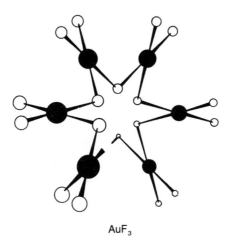

AuF_3

Figure 6.27 Crystal structures of gold(III) chloride and fluoride.

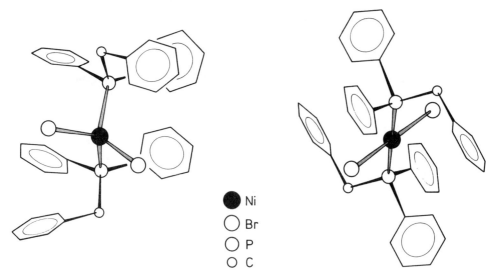

Figure 6.28 In the crystal structure of $NiBr_2 \cdot [P(CH_2C_6H_5)(C_6H_5)_2]_2$, nickel has both a tetrahedral and a square planar geometry.

FIVE COORDINATION

AX_5 Trigonal Bipyramidal and Square Pyramidal Geometries

Trigonal bipyramidal geometry is expected for the spherical d subshells d^0, d^5, and d^{10}. It is also expected for other d subshells if the interaction between the d subshell and the valence shell is weak or if the d subshell has an oblate ellipsoidal shape, in which case the difference between the axial and equatorial bond lengths that is found in compounds of the main-group elements is predicted to be reduced or even reversed (Figure 6.2). If the d subshell has a prolate ellipsoidal shape, then a square pyramidal geometry is predicted in which the axial bond should be longer than the bonds in the base of the square pyramid. Thus, if the d subshell is nonspherical, we cannot predict whether the geometry will be trigonal bipyramidal or square pyramidal. But if the geometry is trigonal bipyramidal, then we expect that the difference between the axial and the equatorial bond lengths will be smaller than it is in a main-group trigonal bipyramidal molecule, and it may be zero or even negative. If the geometry is square pyramidal, then the axial bond is expected to be longer than the four bonds in the base of the square pyramid, rather than shorter than these bonds, as in main-group AX_5E molecules. However, the energy of the square pyramid is only slightly higher than that of the trigonal bipyramid. Thus, even for five equivalent ligands, small effects that are not important in determining other geometries may lead to a preference for a square pyramid geometry, even when a trigonal bipyramid geometry is predicted. For example, molecules in which there are two bidentate ligands often have the square pyramidal geometry even for a spherical d subshell, presumably because the bite of the chelating ligand is better accommodated in the square pyramid geometry. Because of the uncertain effect that multidentate ligands may have on the geometry of a molecule in which the transition metal is five coordinated, we will restrict our discussion mainly to compounds containing only monodentate ligands.

Some examples of trigonal bipyramidal molecules in which the transition metal has a spherical d subshell are given in Table 6.9. In each case the axial bonds are longer than the equatorial bonds, as in the phosphorus pentahalides.

In crystalline $KVO_3.H_2O$ and in V_2O_5 (Figure 6.29), vanadium has a trigonal bipyramidal geometry somewhat distorted by the double-bond character of the nonbridging oxygens, which occupy the equatorial positions. But five-

TABLE 6.9 BOND LENGTHS IN TRIGONAL BIPYRAMIDAL MOLECULES OF d^0, d^5, AND d^{10} TRANSITION METALS

	Molecule	r_{ax} (pm)	r_{eq} (pm)
d^0	VF_5	173.2(7)	170.4(5)
	$NbCl_5$	233.8(6)	224.1(4)
	$TaCl_5$	236.9(4)	222.7(4)
	$TaBr_5$	247.3(8)	241.2(4)
d^5	$FeCl_5^{2-}$		
	$Fe(N_3)_5^{2-}$		
d^{10}	$CdCl_5^{3-}$		
	$Zn(N\text{-methylsalicaldimine})_2$	211	202

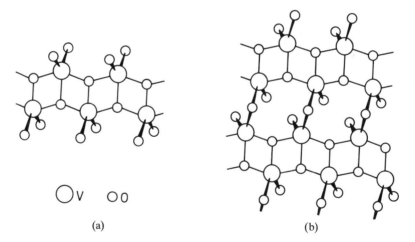

○ V ○ O

(a)   (b)

Figure 6.29 Crystal structures of (a) $KVO_3.H_2O$ and (b) V_2O_5.

coordinated oxoanions of some other transition metals appear to be built from square pyramids; an example is the chains of K_2ZrO_3 (Figure 6.30). The gaseous oxohalides of molybdenum and tungsten, $OMoX_4$ and OWX_4 similarly have square pyramidal structures with the oxygen in the apical position (Figure 6.31). Bond lengths and angles are given in Table 6.10. These molecules would have been predicted to have a trigonal bipyramidal geometry with the oxygen in an

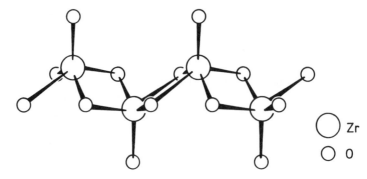

Figure 6.30 In the chain structure of K_2ZrO_3, each zirconium has a square pyramidal AX_5 geometry.

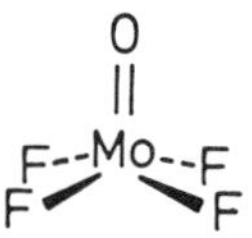

$$\begin{array}{c} O \\ \parallel \\ F\text{--}Mo\text{--}F \\ F \qquad F \end{array}$$

Figure 6.31 Square pyramidal structure of $MoOF_4$.

TABLE 6.10 BOND LENGTHS (pm) AND BOND ANGLES (°) IN GASEOUS $OMoF_4$ AND ANALOGS

OAF_4	$OMoF_4$	OWF_4	$XWCl_4$	$OWCl_4$	$SWCl_4$	$SeWCl_4$
AO	165.0(7)	166.6(7)	WX	168(2)	208.6(4)	220.3(4)
AF	183.6(3)	184.7(2)	WCl	228.0(3)	227.7(3)	228.4(3)
OAF	103.6(3)	104.8(6)	XWCl	102(1)	104.2(5)	104.4(3)
FAF	86.7(3)	86.2(3)	ClWCl	87.3(5)	86.5(2)	86.5(2)

equatorial position, as in SOF_4. The distortion caused by the double bond, which increases the OAF angles to values greater than 90° and 120°, changes the trigonal bipyramid toward a square pyramid and further decreases the small difference in energy between the square pyramid and the trigonal bipyramid, and it is not unreasonable that in some cases the square pyramid becomes the preferred geometry. Moreover, it is probable that there is some interaction between the bonding electrons and the core, as in $VOCl_3$, which would also favor the square pyramidal geometry.

Among the five-coordinated complexes of metals with a d^5 configuration, $Fe(N_3)_5^{2-}$, $FeCl_5^{2-}$, and the dimeric N-methylsalicaldimine complex of Mn(II) have the expected trigonal bipyramidal geometry (Table 6.9). However, $Mn(acac)_2.H_2O$ has a structure that is closer to a square pyramid. No doubt the geometry of the chelating ligand is important in determining whether the geometry is trigonal bipyramidal or square pyramidal in cases like this.

Some five-coordinated zinc and cadmium complexes in which the metal has a d^{10} configuration also have the expected trigonal bipyramidal geometry. Examples include the $CdCl_5^{3-}$ ion and the *N*-methylsalicaldimine complex of Zn(II) (Figure 6.32).

Some examples of molecules of the transition metals with a nonspherical d subshell that have trigonal bipyramidal geometry around the metal atom are listed in Table 6.11. In each case either no difference was observed between the axial and equatorial bond lengths or the axial bonds are shorter than the equatorial bonds as a consequence of the interaction with the d subshell.

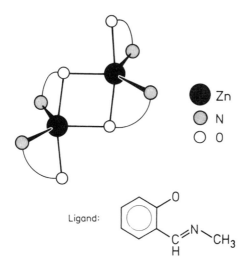

● Zn
◐ N
○ O

Ligand:

Figure 6.32 N-methylsalicaldimine complex of Zn(II).

TABLE 6.11 BOND LENGTHS IN TRIGONAL BIPYRAMIDAL MOLECULES WITH NONSPHERICAL d SUBSHELLS

	Molecule	r_{ax} (pm)	r_{eq} (pm)
d^8	$Pt(SnCl_3)_5^{3-}$	254	254
	$Ni(CN)_5^{3-}$	184	199
	$Co(CNCH_3)_5^+$	187	187
	$Fe(CO)_5$	181	183
d^9	$CuCl_5^{3-}$	230	239
	$[Cu(bipy)_2I]^+$	202	202

Iron pentacarbonyl, $Fe(CO)_5$, in which iron has a d^8 configuration, has a trigonal bipyramidal geometry with shorter axial than equatorial bonds (Figure 6.33). Cobalt pentacarbonyl, $Co(CO)_5$, is not known, but its derivatives, such as $Co(CO)_4SiH_3$ and $Co(CO)_4SiCl_3$, have trigonal bipyramidal geometries with the silyl group in an axial position (Figure 6.33). This geometry is consistent with the Co—CO bonds having some double-bond character and therefore preferentially occupying the equatorial positions. The equatorial Co—CO bonds are tilted toward the longer weaker Co—Si bond and away from the shorter and stronger

Co—CO bond in the axial position. Related molecules include $CoCl(CO)_2[P(CH_3)_3]_2$, $Co_2(CO)_6[P(CH_3)_3]_2$, and $Zn[Co(CO)_4]_2$ (Figure 6.33).

Figure 6.34 shows a nickel(II) complex with a "tripod" tetradentate ligand, which would appear to have a particularly suitable geometry for stabilizing the trigonal bipyramidal geometry at the metal, even when the square pyramid geometry might otherwise be preferred.

There are several examples of molecules of the type $MX_3.2D$, where X is a halogen, D is a donor molecule, and M is a metal with a nonspherical d subshell, that have a trigonal bipyramidal geometry. Examples include $TiBr_3(NMe_3)_2$, $VCl_3(NMe_3)_2$, $VOCl_2(NMe_3)_2$, $VCl_3(SMe_2)_2$, $CrCl_3(NMe_3)_2$, and $NiBr_3(PPh_3)_2$ (Figure 6.35). But again we can find examples of similar molecules, such as

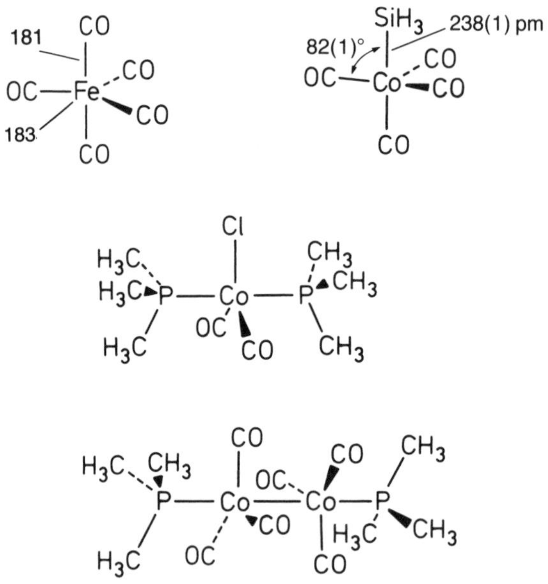

Figure 6.33 Carbonyl compounds of iron and cobalt in which the metal has an AX_5 trigonal bipyramidal geometry.

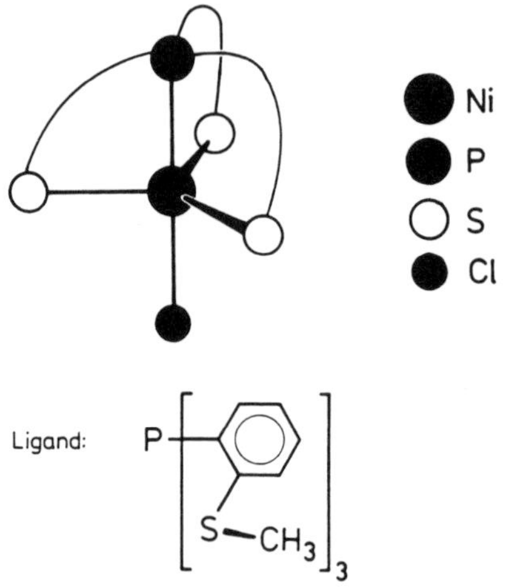

Figure 6.34 Trigonal bipyramidal structure of a Ni(II) complex ion.

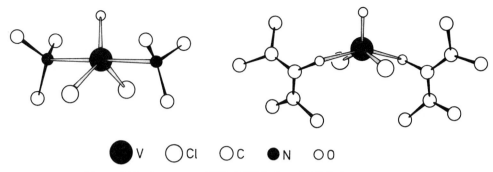

Figure 6.35 Structures of VOCl₂(NMe₃)₂ and VOCl₂(tetramethylurea)₂.

$VOCl_2(tetramethylurea)_2$ (Figure 6.35), that have a geometry that is closer to a square pyramid.

Some examples of molecules of transition metals with a nonspherical d subshell that have a square pyramidal geometry are given in Figure 6.36. In every

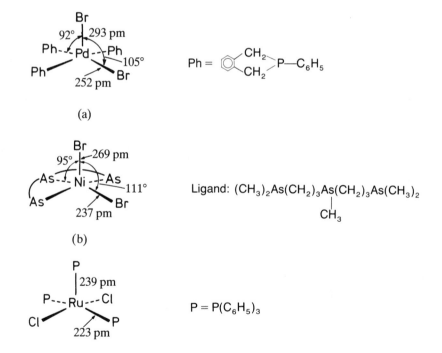

Figure 6.36 Square pyramidal molecules of some transition metals with nonspherical d subshells.
(a) $Pd(II)Br_2[C_6H_4(CH_2)_2PC_6H_5]_3$,
(b) $Ni(II)Br_2[(CH_3)_2As(CH_2)]_3AsCH_3$,
(c) Dichlorotris(triphenylphosphine)ruthenium(II).

case where there are similar bonds in the axial and at least one of the basal positions, the axial bond is longer than the basal bonds, as predicted.

The $Ni(CN)_5^{3-}$ ion in which Ni has a d^8 configuration occurs in both a square pyramidal form and in a trigonal bipyramidal form in $[Cr(en)_3][Ni(CN)_5]1.5H_2O$ (Figure 6.37). In the square pyramidal ion the axial bond (217 pm) is longer than the bonds in the base (186 pm), and in the trigonal bipyramidal ion the axial

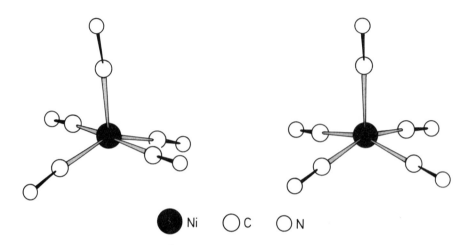

● Ni   ○ C   ○ N

Figure 6.37 The $Ni(CN)_5^{3-}$ ion occurs in both a trigonal bipyramidal and a square pyramidal form in $[Cr(en)_3][Ni(CN)_5]1.5H_2O$.

bonds (184 pm) are shorter than the equatorial bonds (194 pm), as predicted for a nonspherical d subshell.

Many copper(II) chelates form dimers and higher polymers in which copper has square pyramidal geometry. An example of a dimer of this type is given in Figure 6.38.

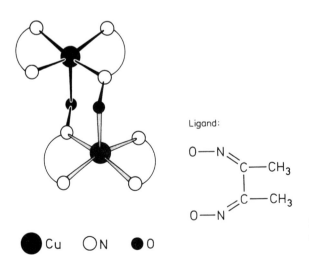

Ligand:

● Cu   ○ N   ● O

Figure 6.38 Dimeric Cu(II) dimethylglyoxine in which Cu has a square pyramidal geometry.

SIX COORDINATION

AX_6 Octahedral and Square Bipyramidal Geometry

Octahedral and distorted octahedral (square bipyramidal) are the most common geometries among the compounds of the transition metals. Molecules in which the transition metal has a spherical d subshell, d^0, d^5, or d^{10}, are expected to have octahedral geometry. If the d subshell is nonspherical and if the interaction between the valence shell and the d subshell is strong enough, then a square bipyramidal geometry is expected (Figure 6.2). This may result either from an elongation or a compression of the octahedron, and it is not possible to predict which type of distortion will be found in any given case.

Examples of octahedral molecules are given in Table 6.12. This table includes both molecules with spherical d^0, d^5 and d^{10} subshells that can be predicted with certainty to be octahedral and also those with nonspherical d

TABLE 6.12   EXAMPLES OF AX_6 OCTAHEDRAL GEOMETRIES

d Subshell	Compound
d^0	$CrF_6(g)$, $MoF_6(g)$, $WF_6(g)$, TiF_6^{2-}, TiO_2, VF_5, $ZrCl_4$, $NbCl_5 \cdot POCl_3$, Nb_2Cl_{10}
d^1	TiF_6^{3-}, $Ti(H_2O)_6^{3+}$, K_2VCl_6, $K_2[CrOCl_5]$, $ZrCl_3$, HfI_3, $NbCl_4$, $TaCl_4$, ReO_3, ReF_6
d^2	$TiCl_2$, VF_3, $V(NH_3)_6^{3+}$, K_2CrF_6, Re_2Cl_{10}, RuF_6, OsF_6
d^3	$V(H_2O)_6^{2+}$, $Cr(NH_3)_6^{3+}$, MnO_2, $ReCl_4$, $KRuF_6$, $NaOsF_6$, IrF_6, CrF_3
d^4	$Mn[S_2CN(CH_3)_2]_3$, K_2OsCl_6, RuO_2, $CsIrF_6$
d^5	$Mn(H_2O)_6^{2+}$, Fe_2O_3, CoF_6^{2-}, K_3RuF_6, IrO_2
d^6	$K_5[Mn(CN)_6]$, $Fe(H_2O)_6^{2+}$, $Fe(CN)_6^{4-}$, CoF_3, K_2NiF_6, $W(CO)_6$, $K_5[Re(CN)_6]$, $PtCl_4(C_5H_{14}N_2)$, AuF_5
d^7	$CoCl_2$, $Co(NH_3)_6^{2+}$, Cs_2CuF_6, $Re_2(CO)_{10}$, $Co(NO_2)_6^{4-}$
d^8	NiO, $KNiF_3$, K_3CuF_6, $Ni(NH_3)_4(NO_2)_2$
d^{10}	AgF, $AgCl$, $AgBr$, $Zn(NH_3)_6^{2+}$

subshells that are nevertheless octahedral because the interaction between the valence shell and the d subshell is not strong enough to distort the octahedral arrangement of the bonding pairs. In the following paragraphs we describe some representative examples of molecules of this type in a little more detail.

The $[Ti(OCH_3)_4]_4$ and $[Ti(OC_2H_5)_4]_4$ tetramers have a cage structure consisting of four connected octahedra (Figure 6.39).

In the solid state the chlorides, bromides, and iodides of niobium, tantalum, tungsten, rhenium, and osmium, and gold pentafluoride in the vapor, exist as halogen-bridged dimers, in which the metal atoms are octahedrally coordinated (Figure 6.40). The pentafluorides of Nb, Ta, Mo, W, Ru, Os, Rh, Ir, and Pt, on the other hand, are tetramers in the crystal (Figure 6.40). The vapors of NbF_5, TaF_5, and MoF_5 contain predominantly trimers with similar halogen-bridged structures (Figure 6.40).

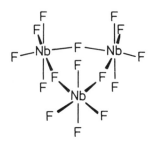

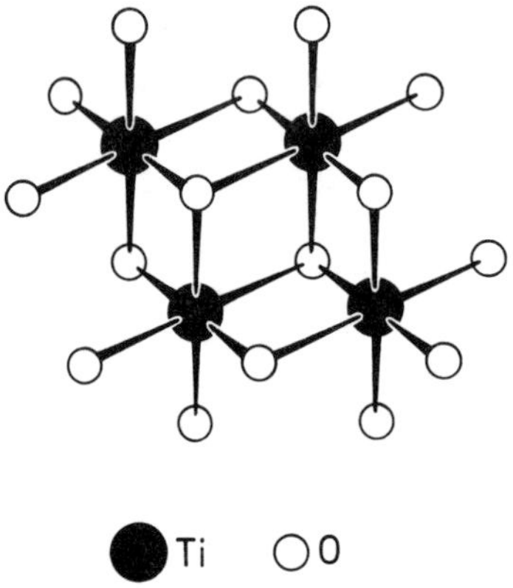

Figure 6.39 Tetrameric structure of titanium methoxide and ethoxide.

Figure 6.40 AuF$_5$ has a dimeric structure in the vapor state. NbF$_5$ is tetrameric in the solid state but trimeric in the vapor.

    In all these dimeric, trimeric, and tetrameric molecules, the bridging bonds are longer than the terminal bonds by about 20 to 30 pm, and the axial bonds are tilted toward the ring and away from the terminal bonds. The deviations from the regular octahedral structure around the metal are consistent with the bridging bonds being longer and weaker than the terminal bonds and, accordingly, exerting weaker repulsions. The weakness of the bridging bonds may be considered to originate from the formal positive charge carried by the bridging halogen, which increases its effective electronegativity, thus decreasing the size of the bonding domain on the central atom.

    Other pentahalides, oxohalides, and complex halides form infinite chains of octahedra, sharing *cis* halogens (VF$_5$, CrF$_5$, ReF$_5$, MoOF$_4$) or *trans* halogens (WOCl$_4$) in the solid state. Many tetrahalides have infinite-chain structures in which, for example, opposite edges of an octahedron (NbI$_4$, α-ReCl$_4$, NbCl$_4$, TaCl$_4$, WCl$_4$, OsCl$_4$) or adjacent edges (TcCl$_4$) are shared. In β-ReCl$_4$ the octahedra share faces and vertices, giving zigzag chains that can be considered to be built up from Re$_2$Cl$_9$ units (Figure 6.41).

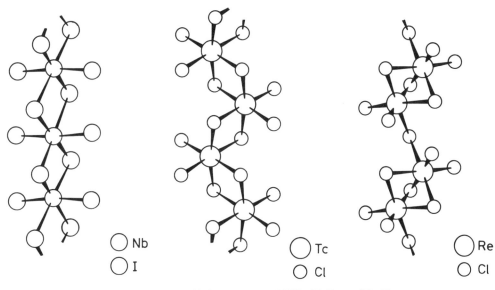

Figure 6.41 Chain structures of NbI$_4$, TcCl$_4$, and ReCl$_4$.

NbOCl$_3$, NbOBr$_3$, MoOCl$_3$, MoOBr$_3$, WOCl$_3$, and WOBr$_3$ have double-chain structures (Figure 6.42). The unit of such a chain is analogous to the M$_2$X$_{10}$ dimers (Figure 6.40), except that the axial halogens are replaced by oxygens, and it is these oxygen atoms that connect these units into chains.

K$_2$NiF$_4$ has a layer structure based on NiF$_6$ octahedra sharing all four equatorial vertices (Figure 6.43). Other complex fluorides and oxofluorides, such as K$_2$ZnF$_4$, K$_2$CoF$_4$, K$_2$NbO$_3$F, and Sr$_2$FeO$_3$F, have similar structures.

Crystalline ReO$_3$ consists of ReO$_6$ octahedra sharing all vertices (Figure 6.44). Another typical crystal structure is rutile, TiO$_2$, in which the titanium atoms are surrounded by six oxygen atoms in an octahedral arrangement (Figure 6.45). Many other transition-metal dioxides adopt this rutile structure.

In those cases when the octahedron is distorted by interaction with a nonspherical d subshell to give a square bipyramid, elongation is more common than compression (Table 6.13). CrF$_2$ and CuF$_2$ have distorted rutile structures

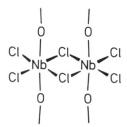

Figure 6.42 Double chain structure of NbOCl$_3$.

Figure 6.43 Layer of NiF$_6$ octahedra sharing corners in the crystal structure of K$_2$NiF$_4$.

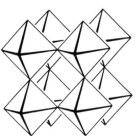

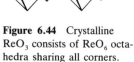

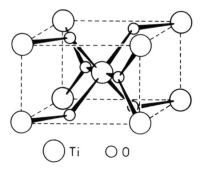

◯ Ti   ◌ O

Figure 6.44 Crystalline
ReO₃ consists of ReO₆ octahedra sharing all corners.

Figure 6.45 Structure of rutile, TiO₂.

TABLE 6.13 BOND LENGTHS (pm) IN AX₆ TETRAGONAL
BIPYRAMIDAL GEOMETRIES

d Subshell	Compound	Equatorial Bonds	Axial Bonds
Elongated octahedra:			
d⁴	CrF₂	200	243
	CrCl₂	239	291
	CrBr₂	254	300
	CrI₂	274	324
	CrS	245	288
	MnF₃	185	209
	K₂MnF₅.H₂O	183	207
	Na₂MnF₅	185	211
d⁹	CuF₂	193	227
	CuCl₂	230	295
	CuBr₂	240	318
	Na₂CuF₄	191	237
	KCuCl₃	229	303
	(NH₄)₂CuCl₄	231	279
	Ba₂[Cu(OH)₆]	197	281
	K₂SrCu(NO₂)₆	204	231
	K₂CaCu(NO₂)₆	205	231
Compressed octahedra:			
d⁴	KCrF₃	214	200
d⁹	KCuF₃	207	196
	K₂CuF₄	208	195
	Rb₂PbCu(NO₂)₆	217	206

with two longer and four shorter metal–fluorine bonds. Similar distortions occur
for CrCl₂ and CuCl₂ (Figure 6.46). Elongated octahedra are connected into a
zigzag chain in the crystal of Na₂MnF₅ (Figure 6.47).

   While the hexanitro complexes of nickel and cobalt contain undistorted
M(NO₂)₆ octahedra, their copper analogs provide interesting examples of distor-
tions to give either an elongated (Figure 6.48) or a compressed square bipyramid.
The $Cu(NO_2)_6^{4-}$ anion of K₂SrCu(NO₂)₆ has an elongated square bipyramidal

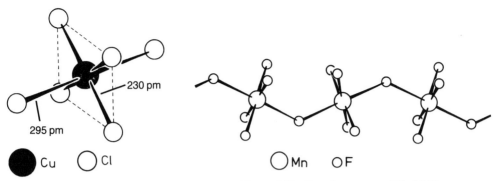

Figure 6.46 Elongated "octahedral" geometry of copper in CuCl₂.

Figure 6.47 Crystal structure of Na₂MnF₅.

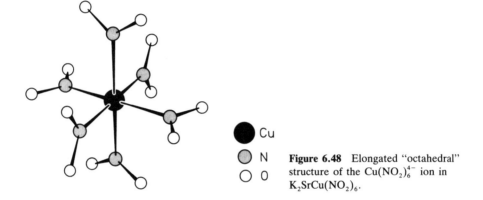

Figure 6.48 Elongated "octahedral" structure of the $Cu(NO_2)_6^{4-}$ ion in $K_2SrCu(NO_2)_6$.

geometry (Figure 6.48), but in the complexes $Rb_2PbCu(NO_2)_6$ and $K_2PbCu(NO_2)_6$ the anion has a compressed square bipyramidal geometry.

Most of the transition-metal dioxo complexes of the type AO_2X_4 have d^0 or d^2 subshells. All the d^0 complexes such as $MoO_2Cl_4^{2-}$ and $VO_2F_4^{3-}$ are *cis*, whereas all the d^2 complexes such as $OsO_2(CN)_4^{2-}$ and $ReO_2py_4^{+}$ are *trans*. According to the VSEPR model, we would expect that all compounds of this type would be *trans* to minimize the interaction between the large double-bond domains.

AX₆ Trigonal Prismatic Geometry

A few six-coordinated complexes of transition metals have a trigonal prismatic geometry. These are almost all dithiolene and related complexes, such as $Re(S_2C_2Ph_2)_3$ and $Mo[Se_2C_2(CF_3)_2]_3$ (Figure 6.49). This geometry does not arise from the interaction of the bonding electron pairs with an ellipsoidal d subshell. Because the S ... S and Se ... Se distances in these compounds are relatively short (300 to 310 pm and 330 pm, respectively), it seems probable that there is bonding between the sulfur (or selenium) atoms to form a trigonal prismatic cage around the metal atom.

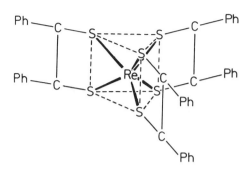

Figure 6.49 Trigonal prismatic geometry of rhenium in the complex $Re(S_2C_2Ph_2)_3$.

SEVEN, EIGHT, AND NINE COORDINATION

Seven Coordination

All three seven-coordinated geometries described in Chapter 3 (Figure 3.22) have been observed among the complex fluorides and oxofluorides of niobium, tantalum, zirconium, and hafnium (Table 6.14).

TABLE 6.14 SEVEN-COORDINATED MOLECULES AND IONS OF THE TRANSITION METALS

Pentagonal Bipyramid	Monocapped Trigonal Prism	Monocapped Octahedron
ZrF_7^{3-}	NbF_7^{2-}	$NbOF_6^{3-}$
HfF_7^{3-}	TaF_7^{2-}	

Eight Coordination

The known eight-coordinated molecules of the transition metals listed in Table 6.15 have either the square antiprism or the dodecahedral geometries discussed in Chapter 3 (Figure 3.23).

TABLE 6.15 EIGHT-COORDINATED MOLECULES AND IONS OF THE TRANSITION METALS

Square Antiprism	Dodecahedron
$Zr(IO_3)_4$	$Zr(C_2O_4)_4^{4-}$
TaF_8^{3-}	$Ti(NO_3)_4$
ReF_8^-	$Mo(CN)_8^{2-}$
$W(CN)_8^{2-}$	ZrF_8^{4-}
$Zr(acac)_4$	

Nine Coordination

The only known nine-coordinated molecules of the transition metals are the two hydride anions ReH_9^{2-} and TcH_9^{2-}, which have the predicted tricapped trigonal prism geometry (Figure 3.24).

REFERENCES AND SUGGESTED READING

F. A. COTTON and G. WILKINSON, *Advanced Inorganic Chemistry*, 5th Ed., Wiley, New York, 1988.

N. N. GREENWOOD and A. EARNSHAW, *Chemistry of the Elements*, Pergamon Press, Oxford, England, 1986.

I. HARGITTAI and M. HARGITTAI, Eds., *Stereochemical Applications of Gas-Phase Electron Diffraction, Part B. Structural Information for Selected Classes of Compounds*, VCH, New York, 1988.

LANDOLT-BÖRNSTEIN, New Series, Volumes II/7 and II/15, *Structure Data of Free Poly-atomic Molecules*, Springer, Berlin, 1976 and 1987.

A. F. WELLS, *Structural Inorganic Chemistry*, 5th Ed., Oxford, University Press, Oxford, England, 1984.

_7_

The Quantum Mechanical Basis of the VSEPR Model

The VSEPR model has been presented in the preceding chapters largely as an empirical extension of the Lewis model. Lewis's suggestion that the electrons in a valence shell can be regarded as arranged in pairs and that a covalent bond consists of a pair of electrons shared between two atoms has proved to be one of the most important and useful ideas in chemistry. It cleared up an enormous amount of confusion concerning valence and bonding and it is still the simplest and most widely used model for discussing bonding, structure, and reactivity. The VSEPR model is an extension of the Lewis model, which proposes that the electron pairs in a valence shell have a definite geometric arrangement from which the arrangement of the bonds around a central atom and hence the geometry of a molecule may be deduced.

Despite the great success of the Lewis model, it is a formal description that tells us nothing about what the sharing of an electron pair really means and how it is that an electron pair can hold two nuclei together. Moreover, it is a static model that takes no account of the motion of the electrons. With the advent of quantum mechanics it became possible, at least in principle, to obtain a more fundamental understanding of the Lewis model and of the nature of the chemical bond in general. In this chapter we review the quantum mechanical description of chemical bonding and we show how it provides a basis for understanding both the Lewis and VSEPR models.

192

QUANTUM MECHANICAL BONDING MODELS

Atomic Orbitals

According to quantum mechanics, the electron in a one-electron atom may be regarded as a three-dimensional standing wave. The form of the wave depends on the energy of the electron and each quantized energy level has associated with it a characteristic standing wave pattern, which is described by a function ψ called the *wave function*. A single-electron standing wave is called an *orbital*. An electron described by the wave function ψ is said to occupy the orbital ψ. The value of $\psi^2 d\tau$ at any given point in space gives the probability that the electron will be found in a very small volume $d\tau = dx\,dy\,dz$ surrounding the point. An alternative but very useful interpretation of ψ is to consider the electron as distributed through space as a *charge cloud*, the density, ρ, of the cloud at any point being proportional to ψ^2. In places where ρ has its largest value, the charge cloud is most dense and more of the negative charge is found. The form of the standing waves or orbitals for the electron in a hydrogen atom or any atom with only one electron (a hydrogenlike atom) can be obtained by solving the Schrödinger equation:

$$\nabla^2 \psi + \frac{2m}{\hbar^2}(E - V)\psi = 0$$

where m is the mass of the electron, E its total energy, V its potential energy, and ∇^2 is the operator

$$\frac{\partial^2}{\partial x^2} + \frac{\partial^2}{\partial y^2} + \frac{\partial^2}{\partial z^2}$$

These standing waves of a hydrogenlike atom are the familiar 1s, 2s, 2p, 3s, 3p, 3d, . . . orbitals.

Although the Schrödinger equation can be solved exactly for one-electron atoms, for a many-electron atom it cannot be solved exactly and it is only an approximation to consider that each electron can be described by a standing wave pattern or that it occupies a given orbital. In order to maintain the useful orbital picture for many-electron atoms, the many-electron Schrödinger equation is broken up into a corresponding number of one-electron equations. An approximate set of atomic orbitals can then be chosen and from these the average potential acting on each electron is calculated. These potentials are then used to calculate new orbitals from which better approximations to the average potential are obtained. The process is repeated until a set of orbitals is obtained that reproduces the potentials that gave these orbitals. Since the orbitals thus obtained resemble hydrogenlike orbitals, they may conveniently be labeled in the same way, 1s, 2s, 2p, 3s . . . . This particular method of obtaining an approximate solution to the many-electron wave function is known as the *self-consistent field method*.

In atoms with two or more electrons, the electrons occupy the orbitals in accordance with the Pauli exclusion principle, according to which identical particles, such as electrons with the same spin, can never be at the same point in space at the same time. This means that each orbital cannot be occupied by more than two electrons that must have opposite spins. The various energy states of the atom correspond to different occupation patterns of the orbitals. In the lowest energy state, or ground state, of an atom, the electrons successively occupy the orbitals from the lowest energy up until all the electrons have been placed in orbitals. This arrangement of electrons among the orbitals in the ground state of an atom is called the electron configuration of the atom in its ground state. For example, the ground state electron configuration of the fluorine atom, which has nine electrons, is written as $1s^2 2s^2 2p_x^2 2p_y^2 2p_z^1$ (Table 1.1).

The Valence-Bond Method

The first quantum mechanical model of a molecule that was developed, the Heitler–London or valence-bond model, is based rather directly on the Lewis model. Electrons are considered to occupy the set of atomic orbitals on each atom. But when a singly occupied orbital on one atom overlaps with a singly occupied orbital on another adjacent atom, and the electrons have opposite spin, either of them can occupy either orbital and thus both electrons are attracted to both nuclei, which decreases their energy. And the greater the orbital overlap is, the greater is the decrease in the energy of the two electrons. In other words, the greater the orbital overlap is, the greater is the energy of the bond formed. This pair of electrons of opposite spin corresponds to the shared pair of electrons in the Lewis model.

When applied to the water molecule, in which the oxygen atom has the valence configuration $2s^2 2p_x^2 2p_y^1 2p_z^1$, maximum overlap is obtained when each H 1s orbital is overlapped along the axis of one of the 2p orbitals. Thus this very simple model predicts that the molecule will be angular but with a bond angle of 90°, rather than the observed angle of 104.5° (Figure 7.1). The model can be improved by considering other possible structures for the water molecule such as

$$\begin{array}{ccc} \mathrm{H^+ \quad O^-} & \mathrm{H{-}O^-} & \mathrm{H^+ \quad O^{2-}} \\ \quad | & & \\ \mathrm{H} & \mathrm{H^+} & \mathrm{H^+} \end{array}$$

and writing the total wave function as a linear combination of the wave functions of each of these structures. By including a sufficiently large number of structures

oxygen
2p orbitals

water molecule   **Figure 7.1**   Valence-bond description
of the water molecule.

the energy and geometry of a molecule can be calculated with considerable accuracy, but the simple physical model of the bonding that we described above that is based on the overlap of directed orbitals in just one structure is lost.

In order to describe the methane molecule in terms of the valence bond method it must first be imagined that the carbon atom is promoted from its ground state $2s^2 2p_x^1 2p_y^1$ in which it would only form two bonds to an excited state in which it has the electron configuration $2s^1 2p_x^1 2p_y^1 2p_z^1$. However, overlapping these orbitals with hydrogen 1s orbitals would predict that in the methane molecule there would be three equivalent bonds at 90° and a fourth nonequivalent bond in some other direction that cannot be specified. In this case a better description of the methane molecule can be obtained by using an alternative set of orbitals called sp^3 hybrid orbitals. Hybrid orbitals are constructed by taking linear combinations of the atomic orbitals. Many different sets of hybrid orbitals can be constructed, but knowing that the four bonds in the methane molecule are equivalent we construct a set of four equivalent orbitals. With an appropriate choice of axes, the wave functions for these orbitals can be written as follows:

$$\psi_{sp^3}(1) = \tfrac{1}{2}(\psi_{2s} + \psi_{2p_x} + \psi_{2p_y} + \psi_{2p_z})$$

$$\psi_{sp^3}(2) = \tfrac{1}{2}(\psi_{2s} + \psi_{2p_x} - \psi_{2p_y} - \psi_{2p_z})$$

$$\psi_{sp^3}(3) = \tfrac{1}{2}(\psi_{2s} - \psi_{2p_x} + \psi_{2p_y} - \psi_{2p_z})$$

$$\psi_{sp^3}(4) = \tfrac{1}{2}(\psi_{2s} - \psi_{2p_x} - \psi_{2p_y} + \psi_{2p_z})$$

They are strongly directed along the tetrahedral directions and each one overlaps a given H 1s orbital more strongly than any one atomic orbital (Figure 7.2). Thus, if we assume that the four bonds in methane are equivalent and that they are therefore formed by four equivalent orbitals, we expect that the methane molecule will have a tetrahedral shape.

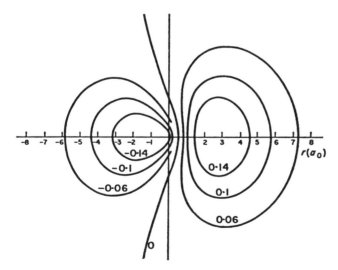

Figure 7.2 Contour diagram of an sp^3 hybrid orbital.

Other sets of hybrid orbitals can be used to describe the bonds in other molecules (Table 7.1). For most molecules we must first construct a set of localized hybrid orbitals that corresponds to the molecular geometry if we wish to

TABLE 7.1   HYBRID ORBITALS

Hybrid Orbitals	Geometry
sp	Linear
sp^2	Equilateral triangular
sp^3	Tetrahedral
sp^{3}d$_{z^2}$^{a}	Trigonal bipyramidal
sp^{3}d$_{x^2-y^2}$^{a}	Square pyramidal
sp^{3}d$_{z^2}$d$_{x^2-y^2}$	Octahedral

^{a}In these cases, because of the nonequivalence of the vertices of these polyhedra there is no unique set of orbitals with this geometry. An infinite number of sets of orbitals can be constructed corresponding to different axial/equatorial bond-length ratios.

use the valence-bond method in which each bond is described by overlapping one singly occupied orbital of one atom with a singly occupied orbital of another atom. For example, to describe the ethene molecule, we use a set of three equivalent sp^2 hybrid orbitals on each carbon atom. The double bond is then described as consisting of a σ bond formed by "end-on" overlap of two sp^2 hybrid orbitals and a π bond formed by "sideways" overlap of two 2p$_z$ orbitals in each carbon atom (Figure 7.3).

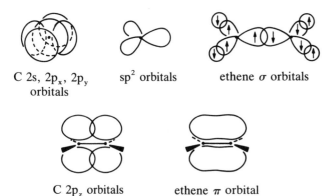

C 2s, 2p$_x$, 2p$_y$ sp^2 orbitals   ethene σ orbitals
orbitals

C 2p$_z$ orbitals   ethene π orbital

Figure 7.3   Valence-bond description of the bonding in ethene.

Hybrid orbitals are used to obtain a simple description of the bonds in a molecule in terms of the valence-bond model, but they do not, in general, provide an a priori explanation for the geometry of a molecule. The hybridization of atomic orbitals is not a phenomenon; it is only a useful mathematical tool for obtaining an approximate description of the bonds in a molecule in terms of the valence-bond method. It is often stated that a carbon atom is sp^3 hybridized, as in methane, or that it is sp^2 hybridized as in ethene, for example, but this is rather loose terminology, because such a statement means only that it is convenient to describe the bonds in terms of sp^3 or sp^2 hybrid orbitals on the carbon atom because the carbon atom is forming four tetrahedral bonds or three trigonal planar bonds, respectively.

The set of four sp^3 hybrid orbitals can also be used to give an alternative description of the water molecule. Two of the hybrid orbitals are singly occupied and are used to form bonds by overlapping with hydrogen 1s orbitals. The other two orbitals are each filled with two electrons and they describe the lone pairs. This description corresponds to a bond angle of 109° for the water molecule (Figure 7.4). The description can be further improved by constructing a set of

Figure 7.4   Valence-bond description of the water molecule based on sp^3 hybrid orbitals.

hybrid orbitals in which the orbitals used for bonding have slightly more p character and the nonbonding orbitals slightly more s character than four equivalent sp^3 hybrids so as to be consistent with the observed bond angle. This description of the bonding in the water molecule serves to emphasize that a set of hybrid orbitals can always be constructed so as to correspond to the geometry of a molecule. Thus the valence-bond method in its simplest form does not enable us to explain the geometry of a molecule; it only enables us to give an approximate description of the nature of the bonds if the geometry of the molecule is known.

For coordination numbers greater than four, the set of hybrid orbitals used to describe the bonding by the valence-bond method must include one or more d orbitals (Table 7.1). In the case of five coordination, one d orbital must be used. If this is the d_{z^2} orbital, an infinite set of $sp^3d_{z^2}$ trigonal bipyramidal orbitals can be obtained, but if the $d_{x^2-y^2}$ orbital is used, a set of square pyramidal orbitals is obtained. Only if the geometry of a five-coordinated molecule is known can a corresponding set of hybrid orbitals be constructed to describe the bonding.

We may conclude therefore that while the valence-bond method at its simplest level is useful for giving a simple approximate description of the bonding in a molecule, it is of little value for the *prediction* of molecular geometry. If the valence-bond method is refined by constructing a linear combination of the wave functions corresponding to each of a large number of structures, we can obtain the energy and geometry of the molecule to a considerable accuracy, but we then no longer have a simple picture of the bonding in the molecule. We will therefore turn to the other important method for describing the electronic structure of molecules, namely, the molecular-orbital method, to see what information it can provide us about molecular geometry.

The Molecular-Orbital Method

The molecular-orbital (MO) theory of the electronic structure of molecules assumes that the electrons in a molecule occupy orbitals that extend over all the nuclei in the molecule, rather than a set of atomic orbitals on each atom. In the lowest energy or ground state of a molecule, the electrons occupy the lowest energy orbitals two at a time in accordance with the Pauli exclusion principle, just as they do when occupying a set of atomic orbitals.

The molecular orbitals for a molecule are usually obtained from a linear combination of atomic orbitals. This is known as the *LCAO approximation*. Near to a nucleus an electron must experience a potential due primarily to that nucleus so that near to a nucleus a molecular orbital must closely resemble an atomic orbital of the appropriate atom. It is therefore assumed that a reasonable approximation for each molecular orbital can be obtained by taking an appropriate linear combination of the atomic orbitals associated with each atom in the molecule. The coefficients in this linear combination are determined by varying the coefficients until the minimum energy is obtained. If a large number of atomic orbitals are included in the expression, a very good approximation to the molecular orbital can be obtained. However, a useful qualitative discussion of the electronic structure of a molecule can often be given in terms of a set of molecular orbitals, each of which is formed from a linear combination of just one atomic orbital on each atom. For example, two molecular orbitals can be constructed from two 1s atomic orbitals as shown in Figure 7.5. The lower energy orbital is called a *bonding orbital* because there is an increased electron density in the internuclear region compared to the individual atoms. The higher energy orbital, which has a node between the nuclei and consequently a decreased electron density in the internuclear region, is called an *antibonding orbital*. This orbital diagram can be used to describe the H_2^+, H_2, and He_2^+ molecules, and it also shows that there is no resultant bonding between two He atoms because both the bonding and antibonding orbitals are filled.

Some information on molecular shape can be obtained from a diagram, which shows the energies of the molecular orbitals for two limiting structures of a molecule and how the orbitals for one structure correlate with those of the other structure. Such a diagram is usually known as a *Walsh correlation diagram*. The approximate relative energies of the orbitals may be estimated by qualitative arguments or they may be calculated to various levels of approximation. Figure 7.6 shows the Walsh correlation diagram for the H_2O molecule and the approximate form of the orbitals for a linear and a bent molecule. The lowest energy orbital, A_1, becomes more stable on bending because of some mixing in of the oxygen $2p_z$ orbital and increased overlap between the H 1s orbitals. For the lowest B_2 orbital the overlap decreases with bending so that its energy increases. The next lowest A_1 orbital is nonbonding in the linear geometry because the orbital overlap is zero but its energy is decreased somewhat as the molecule is bent because of increased overlap. The corresponding B_1 orbital is a pure 2p orbital and therefore is also nonbonding. In this case, there is no change in the overlap with bending so that no change in its energy is expected. However,

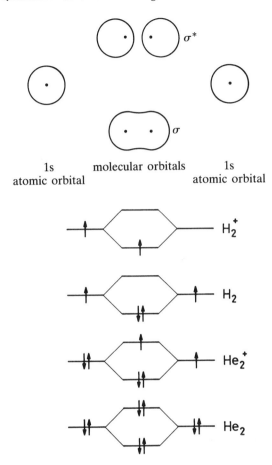

1s
atomic orbital   molecular orbitals   1s
atomic orbital

Figure 7.5 Bonding and antibonding molecular orbitals for the H_2^+, H_2, He_2^+, and "He_2" molecules.

calculation shows that its energy does decrease slightly, probably because of a decrease in the energy of repulsion between the electrons in B_1 and those in the other orbitals. The two highest orbitals, A_1 and B_2, are antibonding. The change in energy of these orbitals is in the opposite direction to that for the two bonding orbitals of the same symmetry. Because the water molecule has only eight electrons, only the four lowest energy orbitals are occupied and so the total energy of the molecule is lower in the bent form than in the linear form.

In principle, the Walsh diagram in Figure 7.6 should have the same general form for any triatomic molecule of the second period elements. But if we were to use the diagram in Figure 7.6 for the BeH_2 molecule, which has four electrons that would occupy only the lowest A_1 and B_2 orbitals, we would predict that the BeH_2 molecule would be bent. However, calculations on BeH_2 show that, on bending, the B_2 orbital increases in energy substantially more than the A_1 decreases so that the linear structure is preferred. In other words, the quantitative aspects of a correlation diagram depend on the specific atoms involved. Only the qualitative features are useful if a single diagram is used for all the members of a family of molecules.

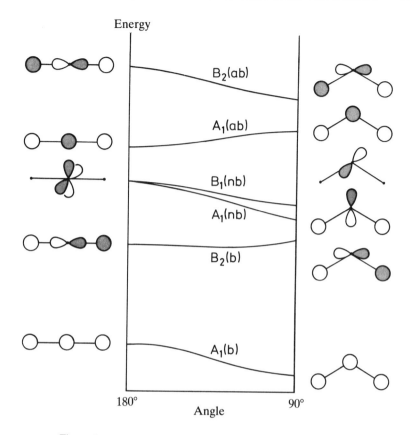

Figure 7.6 Walsh correlation diagram for the H_2O molecule.

Whereas the valence-bond method regards molecules as composed of atoms with pairs of electrons in bonding and nonbonding orbitals on each atom, the molecular-orbital method treats the molecule as a whole, with the electrons occupying molecular orbitals. Thus the description of a molecule in terms of molecular orbitals does not correspond so obviously to the Lewis diagram of the molecule as does the valence-bond description. However, as we see in the next section, molecular orbitals can be transformed to equivalent localized orbitals so as to more nearly correspond to a Lewis diagram.

Equivalent and Localized Orbitals

As we mentioned earlier, the orbitals that are used to describe a many-electron system are not unique. Certain linear combinations of a given set of orbitals describing a molecule form alternative equivalent sets of orbitals that correspond to the same total wave function and have the same total energy and can therefore also be used to describe the molecule. Although the individual orbital densities of one set differ from those of an alternative set, the total density is independent of which orbital set is chosen. In other words, the state of a molecule can be described by any one of a number of equivalent sets of orbitals. Thus the description of a molecule in terms of a given set of orbitals is an arbitrary description. However, a set of orbitals called *canonical orbitals* can be regarded as the most fundamental: they represent the various standing waves of a single electron as if it were moving in the potential field of all the nuclei and the average field due to all the other electrons. The hydrogenlike orbitals normally used to describe atoms are a set of canonical orbitals. However, for some purposes it is convenient to use other alternative equivalent sets of orbitals. *Hybrid orbitals* are a convenient set of equivalent orbitals that can be constructed from the atomic (canonical) orbitals that are useful for the description of bonds in the valence-bond method.

We can similarly transform a set of canonical molecular orbitals into a set of localized orbitals by taking suitable linear combinations. The transformation of the two bonding molecular orbitals in the water molecule to equivalent localized orbitals, one of which is largely localized in the region of one of the OH bonds and the other in the region of the other OH bond, is shown in Figure 7.7. The two

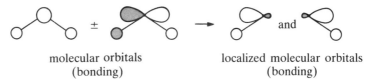

molecular orbitals
(bonding)

localized molecular orbitals
(bonding)

Figure 7.7   Localized molecular (bond pair) orbitals for the water molecule.

largely nonbonding orbitals may also be transformed in the same way to produce two orbitals that correspond to the two approximately tetrahedrally oriented lone pairs of the VSEPR model. This set of localized orbitals is entirely equivalent to the initial set of delocalized molecular orbitals.

Approximate localized molecular orbitals may also be obtained by combining appropriate atomic orbitals on just two adjacent atoms in a molecule. This method corresponds closely to the valence-bond method in which electrons in two singly filled orbitals on adjacent atoms are paired together. For example, if we start with a set of tetrahedral hybrid orbitals on the oxygen atom, we may combine two of them with hydrogen 1s orbitals to form two localized molecular orbitals that correspond to the two O—H bonds, leaving the other two tetrahedral hybrid orbitals filled with two electrons each, which correspond to the lone pairs of the Lewis and VSEPR models.

No matter how we express the orbitals, the simple qualitative LCAO molecular-orbital method is limited in its power to predict molecular geometry. In order to use the molecular-orbital method to obtain reliable information about molecular geometry, it is necessary to use the self-consistent field method or other methods to obtain the geometry with the lowest energy. This type of calculation, which starts with a large basis set of atomic orbitals and in which the geometry is varied until the energy of the molecule is minimized, is known as an *ab initio calculation*. With the aid of modern high-speed computers the geometries of many molecules, particularly those composed of light atoms, have been determined by this method. Agreement with the experimental geometry in those cases where it is known is sufficiently good that we can have reasonable confidence in the calculated geometry of a molecule for which there is no experimental data. However, useful as such calculations are, they do not give us much insight into the reasons why a molecule adopts a given geometry nor do they help us to better understand the Lewis and VSEPR models.

ELECTRON DENSITY DISTRIBUTIONS

One very important property of a molecule that we can obtain from ab initio calculations is the total electron density distribution (Figure 7.8).

The description of the electronic structure of a molecule in terms of orbitals, whether they be molecular, hybrid, or equivalent orbitals, is a useful but nevertheless arbitrary description. An orbital is not a fundamental property of a molecule that can be verified by experiment. However, by finding the sum of the electron densities of each of the individual orbitals, we can obtain the total electron density for the molecule. And the total electron density is an important

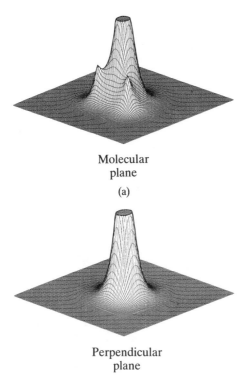

Molecular
plane

(a)

Perpendicular
plane

(b)

Figure 7.8 Plots of the total electron density, ρ, for the water molecule (a) in the molecular plane and (b) in a plane through the oxygen perpendicular to the molecular plane and bisecting the HOH angle. (Reproduced with permission of R. F. W. Bader)

property of a molecule that can also, at least in principle, be obtained by experiment.

We might perhaps expect to see evidence for bonding and nonbonding electron pairs in the total electron density (Figure 7.8), but the only prominent features are the strong peaks at the positions of the nuclei due to the high density of the core electrons. There are no obvious signs of any features corresponding to electron pairs, either bonding or nonbonding. However, this is not too surprising because the electron density changes that occur on molecule formation are very small compared to the large electron densities that surround each nucleus. In the following sections we examine several different ways in which the total electron density can be examined in more detail in order to see what evidence it provides for the bonding and nonbonding electron pairs of the Lewis and VSEPR models. In particular we will consider:

1. Localized molecular orbitals from ab initio calculations
2. Electron density difference maps
3. The Laplacian of the electron density

Localized Molecular Orbitals from ab Initio Calculations

It is only the total electron distribution obtained by summing the densities of the electrons in all the molecular orbitals that has any real physical significance, because the individual orbitals are arbitrary since many different equivalent sets of orbitals are possible that all give the same total electron density distribution. However, let us see what information can be obtained from ab initio calculations if the canonical molecular orbitals are transformed into localized orbitals.

We will take as an example the work of Schmiedekamp et al. (1979). They calculated the canonical set of molecular orbitals and the geometry for a number of simple hydrides, fluorides, and related molecules using the LCAO–SCF method (Table 7.2). They converted the canonical set of orbitals into a localized set by maximizing the sum of the squares of the distances between the centroids of charge of the orbitals. Plots of the electron density in the orbital corresponding

TABLE 7.2 CALCULATED AND EXPERIMENTAL GEOMETRIES

Molecule	Bond Length (pm)		Bond Angle (°)	
	Calculated	Experimental	Calculated	Experimental
OF_2	140.7	140.5	102.0	103.0
SF_2	161.5	158.7	98.3	98.0
SH_2	133.5	133.5	93.6	92.2
SOH_2	136.1	—	89.2	—
SO_2H_2	134.6	—	97.7	—
SO_2	143.4	143.1	118.2	119.3
PH_3	141.4	141.2	94.0	93.4
NH_3	100.8	101.3	105.4	107.4
OH_2	95.0	96.2	104.9	104.6

to one of the sulfur lone pairs, to the S—F bond in SF_2 and to the S—H bond in H_2S are shown in Figure 7.9. It can be seen that the lone-pair orbital is larger and more spread out around the sulfur core than the SF bonding orbital, and the centroid of charge of the lone-pair orbital is closer to the sulfur core than the

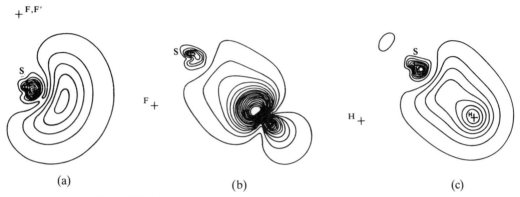

(a)   (b)   (c)

Figure 7.9 Electron density plots for (a) a localized lone-pair orbital in SF_2, (b) a bond-pair orbital in SF_2, and (c) a bond-pair orbital in SH_2. Reproduced with permission of A. Schmiedekamp.

centroid of charge of the S—F bond orbital. By taking the centroid of charge of the lone-pair orbital as the position of the lone pair, the angles between lone pairs and between lone pairs and bond pairs are readily calculated. Some values are given in Table 7.3. In each case, for a particular molecule it may be seen that

TABLE 7.3 COMPUTED ANGLES (°) INVOLVING LONE PAIRS

Molecule	HX(lp)	FX(lp)	(lp)X(lp)
OF_2	—	104.3	133.8
SF_2	—	104.4	135.2
OH_2	108.2	—	118.5
SH_2	107.8	—	126.9

these angles decrease in the order lp:lp > bp:lp > bp:bp in accordance with the predictions of the VSEPR model. Another interesting way to present the results of these calculations is to compare the space requirements of lone pairs, multiple bonds, and single bonds by calculating the average of the three angles made by a given bond or lone pair with its three neighbors. This so-called triple average angle gives a measure of the solid angle required by each bond or lone pair. These triple average angles are given in Table 7.4. They have rather constant values from one molecule to another. The values for lone pairs are always larger than for bonds, the SF bond pair is slightly smaller than the S—H bond pair, which is consistent with the greater electronegativity of fluorine, and the S=O double

TABLE 7.4 TRIPLE AVERAGE ANGLES (°) FOR BONDS AND LONE PAIRS

Bond or Lone Pair	Molecule	Triple Average Angle	Bond or Lone Pair	Molecule	Triple Average Angle
S—F	HSF	102.2	O—H	OH_2	107.1
	SF_2	102.4	O—F	OF_2	103.5
S—H	HSF	103.8	O—(lp)	OF_2	114.1
	SH_2	103.1		OH_2	111.6
	SOH_2	103.1	N—H	NH_3	108.0
	SO_2H_2	104.7		NH_2^-	103.8
S=O	SOH_2	113.6	N—(lp)	NH_3	113.5
	SO_2H_2	113.3		NH_2^-	113.8
S—(lp)	SH_2	114.2	P—H	PH_3	103.5
	HSF	114.2	P—(lp)	PH_3	122.4
	SOH_2	114.8			
	SF_2	114.7			

bond is larger than the SF and SH single bonds. All of these results are consistent with the VSEPR model. We see also that the S=O double bond has almost the same space requirement as a lone pair on sulfur, which is a result that cannot be predicted with certainty on the basis of the VSEPR model, although we suggested in Chapter 5 that for third-row elements such as sulfur and phosphorus it seems probable that lone-pair domains and double-bond domains are of approximately the same size.

Comparison of the angles involving the lone-pair orbitals in AX_2E_2 molecules are also of considerable interest because of the apparently anomalous bond angles in some of these molecules. For example, the bond angle in H_2S is smaller than that in SF_2 despite the greater electronegativity of fluorine. However, reference to Tables 7.3 and 7.4 shows that space requirement of the SF bond pair is, as expected, slightly smaller than that of the SH bond pair. We see also that the lp:lp angle increases from H_2S to HSF to SF_2 as the smaller size of the SF bond pair domain allows the lone pairs to spread farther apart. In each case, also, the SH bond pair makes a larger angle with the lone pair than the smaller SF bond pair. All these results are in accord with the VSEPR model. The bp:bp angle in H_2S appears anomalous because it is smaller than the bp:bp angle in F_2S, but as we have pointed out previously, it is not possible to predict this angle with certainty because the lone pairs are in a plane that is perpendicular to the plane of the bond pairs.

Although the molecular-orbital method in its simpler forms has only a limited usefulness for the prediction of molecular geometry, if the calculations are taken to a sufficient level, the geometry of relatively simple molecules can be obtained with considerable accuracy. But, unfortunately, the nature and complexity of such calculations preclude us from obtaining any understanding of the factors that are important in determining molecular geometry. However, if the canonical molecular orbitals are converted to localized orbitals, insofar as these localized orbitals represent the electron distribution of individual pairs of electrons, they give considerable support to the basic ideas of the VSEPR model.

Unfortunately, for most molecules, the so-called "localized" orbitals still overlap to a considerable extent, and therefore the electrons are not, in fact, well localized into separated pairs. As we will see, it is not in general possible to construct a set of localized orbitals such that each orbital represents even approximately an individual electron pair.

Density Difference Maps

The total electron density distribution of a molecule does not obviously show any of the perturbations in the valence regions of the atoms that might arise from the formation of bonds because the overall appearance of the charge distribution is overwhelmingly dominated by the high charge densities of the atomic cores. One possible way of obtaining further information from the total charge density determined by X-ray crystallography or by ab initio molecular-orbital calculations is to subtract from the total charge density the superimposed spherical charge densities of the isolated atoms placed at the positions of the nuclei in the molecule to give a density difference map. In suitable cases, if the electron density distribution of the molecule is known with a sufficiently high accuracy, a difference density map shows residual electron density corresponding to all the bond pairs and lone pairs that would be expected from the Lewis diagram and with the arrangement predicted by the VSEPR model (Figure 7.10). However, the assumption that the superimposed charge densities of free atoms can be used to construct a comparison "molecule" in which the atoms are not bonded together has been questioned and several alternatives have been proposed, such as the use of nonspherical atom charge densities. The features of the density difference map depend on the method used to construct the comparison "molecule"; consequently, the interpretations resulting from such an analysis must be viewed as subjective and not necessarily indicative of any real physical property of the molecule. Nevertheless, the electron density difference patterns do suggest that the total electron density is indeed distorted from the sum of the free atom density distributions in such a way that there are concentrations of electron density in those regions where bonding pairs and lone pairs are presumed to be located in the Lewis and VSEPR models.

We will see shortly how the presence of such concentrations of electron density in the total electron density can be revealed without having to resort to density difference maps.

Atomic Fragments and Bond Paths

The total electron density distribution has maxima only at the positions of the nuclei and shows no maxima corresponding to bonds or lone pairs, nor does it even clearly distinguish between the electron density distribution of one atom and that of its neighbors (Figure 7.8). It would be very useful to be able to divide up the total electron density into fragments corresponding to each of the atoms in the molecule. But if this is done in some arbitrary manner, these fragments are, in general, not even approximately transferable from one molecule to another. It

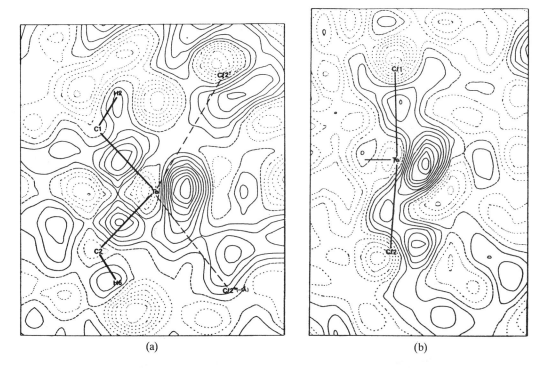

(a)   (b)

Figure 7.10 Electron density difference map for the $(CH_3)_2TeCl_2$ molecule. (a) In the equatorial Te, C1, C2 plane showing the Te—C bond and Te(IV) lone-pair densities. Contours are at $0.03\,e/Å^3$ with negative contours broken. (b) Perpendicular to the plane in (a) showing the Te—Cl bond and Te(IV) lone-pair densities. Reproduced with permission, Ronald F. Ziolo and Jan M. Troup, *J. Amer. Chem. Soc.*, **105**, 229 (1983).

has been shown by Bader and Essen (1984), however, that there is a unique way of partitioning the electron density distribution into fragments so that the charge distribution of each fragment suffers a minimum of change when it is transferred between similar molecules, and the properties of such fragments are additive to a good approximation. The fragment corresponding to a given nucleus is defined by the space occupied by the gradient paths or lines of steepest descent starting from that nucleus. Figure 7.11(a) shows the total electron density for the ethene molecule in the molecular plane. Figure 7.11(b) gives the gradient paths for each of the atoms in the molecular plane and shows how they divide the electron density into fragments corresponding to each of the atoms. This is made still clearer in (c) where the two gradient paths (lines of steepest descent) starting from the point of minimum density or saddle point (●) between each pair of nuclei are also shown. These particular gradient paths form the boundary between each atomic fragment in the molecular plane. The collection of all such gradient paths starting from a saddle point defines the partitioning surface between a pair of atomic fragments. Such a surface is also called a *zero flux surface*. The two lines starting at the saddle point or point of minimum density between a pair of

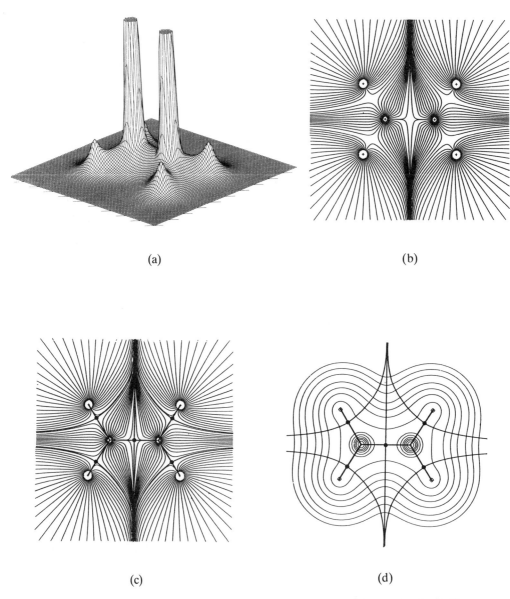

(a)

(b)

(c)

(d)

Figure 7.11   Partitioning surfaces and bond paths in the ethene molecule. (a) The total electron density of the ethene molecule in the molecular plane. (b) Gradient paths (lines of steepest descent) from each nucleus form the basis for the division of the total electron density into fragments corresponding to each of the atoms. (c) The partitioning surface between two atoms is defined by the collection of gradient paths starting at the point of minimum density or saddle point (●) between each pair of atoms. The bond path is formed by the two lines of steepest ascent starting at the saddle point (●) in a direction perpendicular to the partitioning surface. (d) The partitioning surfaces and bond paths are shown superimposed on a contour plot of the electron density in the molecular plane. (Reproduced with permission of R. F. W. Bader)

nuclei in a direction perpendicular to the partitioning surface and taking the path of steepest ascent to each of the adjacent nuclei define a ridge of electron density that is called a *bond path*. The five bond paths in the ethene molecule are also shown in (c). In (d) the bond paths and partitioning surfaces in the molecular plane are shown superimposed on a contour plot of the total electron density. Figures 7.12 and 7.13 show similar contour plots for the H_2O and BF_3 molecules in the molecular plane together with the partitioning surfaces and bond paths.

Thus we can delineate the regions of the electron density belonging to each atom in a molecule and we can recognize the feature of the total electron density that corresponds to the bonds between the atomic fragments. But the overwhelming contribution from the electron densities of the atomic cores apparently prevents us from obtaining any detailed information about the bond pairs and lone pairs that are an essential feature of the Lewis and VSEPR models. However, Bader, MacDougall, and Lau (1984) have shown that the Laplacian of the electron density distribution greatly magnifies the very small perturbations of the total electron density that result from bond formation and thus enables their details to be studied.

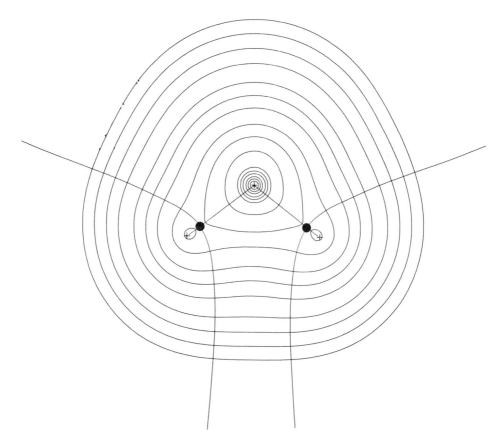

Figure 7.12 Partitioning surfaces and bond paths for the water molecule. (Reproduced with permission of R. F. W. Bader)

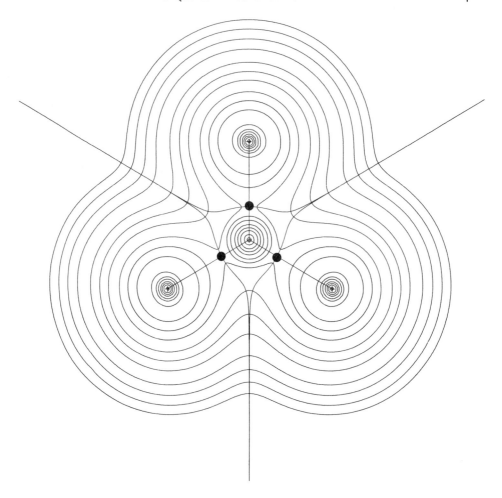

Figure 7.13 Partitioning surfaces and bond paths for the boron trifluoride molecule. (Reproduced with permission of R. F. W. Bader)

THE LAPLACIAN OF THE ELECTRON DENSITY

The Laplacian is the second derivative of a function in three dimensions. To understand how the Laplacian reveals features of the electron density distribution in a molecule that are not readily apparent in the total electron density, we will first consider the Laplacian of the electron density of an atom. A plot of the total electron density of the argon atom in any plane through the nucleus is shown in Figure 7.14. There is only a single peak in the electron density with a maximum at the nucleus. The plot has been cut off at a low value of ρ so that the density in the outer valence shell region can be seen more clearly. But no maximum in the electron density corresponding to the valence shell nor indeed to any of the K, L, and M shells can be seen. The electron density along any radial line from the

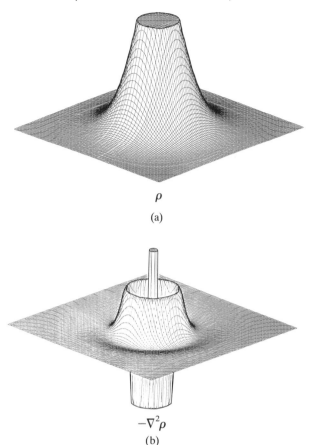

ρ

(a)

$-\nabla^2\rho$

(b)

Figure 7.14 (a) Total electron density of the argon atom in a plane through the nucleus and (b) the negative Laplacian of the electron density in the same plane. (Reproduced with permission of R. F. W. Bader)

nucleus decreases monotonically. Figure 7.15 shows a plot of an arbitrary one-dimensional function $f(x)$ that mimics the radial behavior of the electron density of a second-period element. In this figure we also show the first and second derivatives, $df(x)/dx$ and $d^2f(x)/dx^2$, and the negative of the second derivative $-d^2f(x)/dx^2$. For all values of x, the first derivative $df(x)/dx$ is negative, showing the absence of any maximum in $f(x)$. However, the slope of $f(x)$ does not decrease continuously but reaches a minimum value and then increases slightly and shows a point of inflection at x corresponding to the shoulder in $f(x)$. The second derivative $d^2f(x)/dx^2$ shows a pronounced minimum at this point and therefore $-d^2f(x)/dx^2$ has a maximum at this point. At a point such as x_1 in the function $f(x)$ representing the radial behavior of the total charge density of an atom, we may say that the charge is locally concentrated in the sense that although $f(x)$ is not greater than both $f(x + dx)$ and $f(x - dx)$ because there is no maximum at this point, it is greater than the average of $f(x + dx)$ and $f(x - dx)$. Conversely, the term locally depleted implies that $f(x)$ is less than the average of the values of the function at neighboring points. We see therefore that the slight shoulder in $f(x)$ becomes a pronounced maximum in $-d^2f(x)/dx^2$, thus making the shoulder much more apparent. Similarly,

$$f(x) = 8\exp(-7x) + \exp[-10(x-0.5)^2]$$

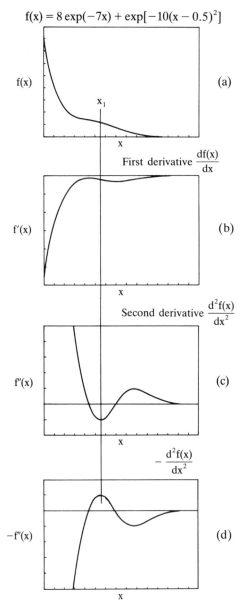

First derivative $\dfrac{df(x)}{dx}$

Second derivative $\dfrac{d^2f(x)}{dx^2}$

$-\dfrac{d^2f(x)}{dx^2}$

Figure 7.15 (a) A one-dimensional function that decreases monotonically with increasing x but has a shoulder at x_1. (b) The first derivative $df(x)/dx$. (c) The second derivative $d^2f(x)/dx^2$. (d) The negative second derivative $-d^2f(x)/dx^2$ exhibits a maximum at the position of the shoulder in $f(x)$. (Reproduced with permission of R. F. W. Bader)

$$-\nabla^2\rho = -\left(\frac{\partial^2\rho}{\partial x^2} + \frac{\partial^2\rho}{\partial y^2} + \frac{\partial^2\rho}{\partial z^2}\right)$$

shows where the three-dimensional charge distribution is locally concentrated and depleted. Figure 7.14 also shows a plot of $-\nabla^2\rho$ for the argon atom which can be compared with the plot of the total electron density. The plot of $-\nabla^2\rho$ clearly shows three maxima or regions of local charge concentration, each of which is surrounded by a region of local charge depletion. These local charge concentrations correspond to the K, L, and M quantum shells of the argon atom. Figure 7.16 shows a section through the nucleus for a similar plot for the krypton atom. Here we see the local charge concentrations corresponding to the K, L, M, and N shells of the krypton atom. These shells are not evident in the total charge density because there are no maxima and minima in this total density, although incorrect diagrams showing such maxima and minima in the total electron density are sometimes given in elementary texts.

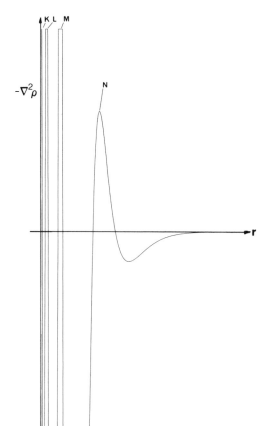

Figure 7.16 A section through the negative Laplacian $-\nabla^2\rho$ of the electron density of the krypton atom showing the four charge concentrations corresponding to the K, L, M, and N shells. (Reproduced with permission of R. F. W. Bader)

A function called the *radial distribution function*, which is the integrated electron density in an infinitesimal spherical shell of radius r and thickness dr, is often used to demonstrate the existence of the electron shells of an atom. This one-dimensional function does show maxima and minima corresponding to the electron shells, but this does not mean that there are maxima and minima in the three-dimensional electron density distribution in real space. Moreover, the radial distribution function is not applicable to atoms in molecules, whereas, as we will see, the shell structure of atoms, as revealed by the Laplacian, persists in molecules.

Valence-shell Charge Concentrations

The outermost shell of charge concentration of an atom is called the *valence-shell charge concentration* (VSCC). An atom in a molecule generally possesses the same number of shells of charge concentration as it does in isolation, except in molecules where the bonding is predominantly ionic, that is, where one atom has lost most of its valence electrons to its bonding partner. In such a case, the VSCC of the element forming the positive ion is eliminated upon molecule formation. In the general case where the VSCC of each atom persists after molecule formation, the VSCC is not spherical and it is not of uniform magnitude, but it does exhibit local maxima and minima. The distortion of the VSCC that accompanies the formation of bonds is illustrated in Figure 7.17. This figure shows the negative Laplacian of the total electron density, $-\nabla^2\rho$, for the water molecule in the plane of the molecule and in a plane through the oxygen atom perpendicular to the plane of the molecule. The valence-shell charge concentration is no longer

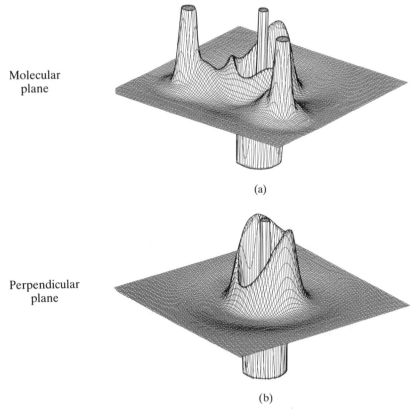

Molecular
plane

(a)

Perpendicular
plane

(b)

Figure 7.17   (a) The negative Laplacian, $-\nabla^2\rho$ of the total electron density for the water molecule in the molecular plane and (b) in a plane through the oxygen perpendicular to the molecular plane and bisecting the HOH angle (cf. Figure 7.8). (Reproduced with permission of R. F. W. Bader)

uniform as it would be in the free oxygen atom, but in the plane of the molecule it exhibits three apparent maxima. Two of the maxima are between the oxygen atom and the hydrogen atoms and they are the local charge concentrations corresponding to each bond: they are called *bonding charge concentrations*. There are also two maxima in the VSCC of the oxygen atom in the plane perpendicular to the plane of the molecule. These two maxima in the VSCC are the local charge concentrations corresponding to the lone pairs in the valence shell of oxygen. What appears to be a third maximum in the plane of the molecule is not a local maximum but is the saddle point between the two maxima in the plane perpendicular to the molecular plane. Thus there are two bonding charge concentrations and two nonbonding charge concentrations in the valence shell of the oxygen atom so that there is a direct correspondence between the number of electron pairs in the valence shell of oxygen in the Lewis diagram of the molecule and the number of valence shell charge concentrations. Moreover, the charge concentrations have the approximately tetrahedral arrangement proposed by the VSEPR model for four electron pairs. The same correspondence between the number and arrangement of the valence-shell charge concentrations of an atom in a molecule and the number and arrangement of the electron pairs in the valence shell of the atom as proposed in the Lewis and VSEPR models has been found in many other molecules. For example, Figure 7.18 shows a contour map of the Laplacian of the electron density for the NH_3 and PH_3 molecules in a plane through one of the bonds and containing the threefold symmetry axis of the molecule. There is a charge concentration in the valence shell of the central atom corresponding to each of the bonds and also a single nonbonding charge concentration. The nonbonding valence-shell charge concentration on the phosphorus atom in PH_3 is larger and more spread out around the phosphorus core than the nonbonding charge concentration on the nitrogen atom and the angle between the PH bond charge concentrations is smaller than that between the NH bond charge concentrations, as expected according to the VSEPR model. The position of each charge concentration is taken to be the position of its maximum. It would be difficult to map the exact shape of each charge concentration on the surface of maximum charge concentration, so the surface area of each charge concentration is approximated by the portion of the surface of a sphere with a radius equal to the distance of the maximum of the charge concentration from the nucleus bounded by a cone subtending an angle α at the nucleus, as determined by the neighboring saddle points. The relative "thickness" of each charge concentration is measured by μ, the radial curvature of $-\nabla^2\rho$.

Information on the positions, sizes, and shapes of the charge concentrations in CH_4, NH_3, and H_2O is given in Table 7.5. In each molecule there are four valence-shell charge concentrations, but the valence shell is no longer spherical. The distance, r, of a bonded charge concentration from the nucleus is increased by 2 to 4 pm over the distance of the sphere of maximum charge concentration from the nucleus in the free atom, while the nonbonding charge concentrations are drawn slightly closer. The nonbonding charge concentrations have a larger area than the bonding charge concentrations and they are thinner in radial extent. Thus a nonbonding charge concentration, while radially more contracted than a

TABLE 7.5 BONDED AND NONBONDED CHARGE CONCENTRATIONS

Bonded Charge Concentrations:

	r (pm)	$-\nabla^2\rho$ (au)	bAb (°)	Area (10^{-2} pm^2)	Thickness (μ) (au)	nAb (°)
CH$_4$	53.7(49.8)[a]	0.717	109.5	91	10.1	—
NH$_3$	43.4(41.1)	1.911	106.6	48	77.7	112.3
OH$_2$	37.4(34.8)	2.688	103.1	32	223.4	102.2
PH$_3$	94.5	0.638	96.1	161	2.5	120.8
PF$_3$		Charge transfer to fluorine				
NF$_3$	47.3(41.1)	0.990	97.6	45	28.2	119.6
SH$_2$	84.1	0.806	95.4	117	4.4	105.4

Nonbonded Charge Concentrations:

	r (pm)	$-\nabla^2\rho$ (au)	nAn (°)	Area (10^{-2} pm^2)	Thickness (μ) (au)	Net Charge on H
CH$_4$	—	—	—	—	—	−0.06
NH$_3$	38.8(41.1)	3.210	—	74	236.5	+0.37
OH$_2$	33.5(34.8)	6.616	138.3	45	715.1	+0.62
PH$_3$	77.5	0.325	—	249	7.8	−0.63
PF$_3$	75.0	0.502	119.8	215	12.1	−0.90
NF$_3$	35.8(41.1)	4.450	119.6	74	294.7	−0.36
SH$_2$	69.6	0.554	133.7	220	17.3	−0.34

[a]Values in parentheses are for the free atom.

bonding charge concentration, is laterally more diffuse and possesses a larger surface area. As the electronegativity of the central atom increases from carbon to nitrogen to oxygen, the bonding charge concentration moves closer to the nucleus of the central atom and decreases in area. Because a nonbonding charge concentration is larger than a bonding charge concentration in the same valence shell, the angle between a bonding charge concentration and a nonbonding charge concentration, nAb, is larger than the tetrahedral angle, while the bAb angle is smaller than the tetrahedral angle in NH$_3$. In H$_2$O the angle between the nonbonding charge concentrations is 138° and is considerably larger than either the nAb or bAb angles.

On moving from second-row to third-row elements (Table 7.5), the considerably increased size of the core in H$_2$S and PH$_3$ compared to H$_2$O and NH$_3$ is reflected in the much larger area of the nonbonding charge concentration. The increased size of the nonbonding charge concentration causes a corresponding decrease in the angle between the bonding charge concentrations and therefore in the bond angles in H$_2$S and PH$_3$, which are smaller than in H$_2$O and NH$_3$. Even though the bond angles are considerably smaller, because of the larger size of the phosphorus and sulfur cores the bonding charge concentrations are actually farther apart in H$_2$S than in H$_2$O (120 pm compared with 68 pm) and in PH$_3$ than in NH$_3$ (141 pm compared with 81 pm).

On comparing PF$_3$ with PH$_3$ (Figure 7.18), we see that there are no longer any bonding charge concentrations in the valence shell of phosphorus, although there is a nonbonding charge concentration. The bonding charge concentrations have been transferred to the valence shell of the fluorine, which is essentially a fluoride ion with its charge density polarized toward the positively charged

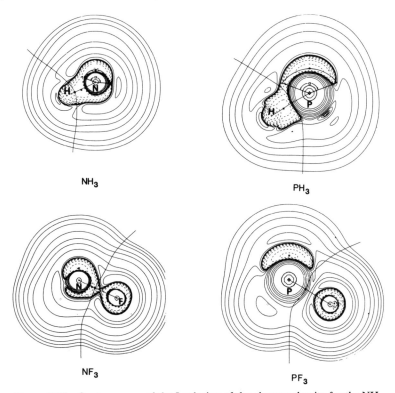

NH₃

PH₃

NF₃

PF₃

Figure 7.18   Contour maps of the Laplacian of the electron density for the NH_3, NF_3, PH_3, and PF_3 molecules. Dotted lines show where the Laplacian is negative corresponding to a local concentration of charge. (Reproduced with permission of R. F. W. Bader)

phosphorus core. By summing the total electron density within the boundaries of the fluorine atom, the charge on each fluorine is found to be -0.90, leaving a charge on phosphorus of $+2.70$.

Valence Shells with Five Charge Concentrations

We saw in Chapter 5 that the bond angles and bond lengths in molecules whose shapes are based on the trigonal bipyramidal arrangement of five electron pairs provide excellent examples of all the effects predicted by the VSEPR model. It is of particular interest therefore to examine the properties of the valence-shell charge concentrations in molecules of this type.

Table 7.6 summarizes the properties of the valence-shell charge concentrations for ClF_3, SF_4, and OSF_4. In all three cases the minimum energy geometry was found by ab initio calculations to be that of a trigonal bipyramid, and five charge concentrations are found in the valence shell of the central atom in each case. Figure 7.19 shows the negative Laplacian of the electron density in the equatorial plane and in the plane of the ClF_3 molecule. In the equatorial plane there are one bonding charge concentration and two nonbonding charge

TABLE 7.6 BONDED AND NONBONDED CHARGE CONCENTRATIONS IN ClF_3, SF_4, SOF_4, and ClF_5

Bonded Charge Concentrations:

Molecule		r (pm)	$-\nabla^2\rho$ (au)	Area $(10^{-2}\,pm^2)$	Thickness (μ) (au)	b_aAb_e (°)	nAb_a (°)
ClF_3	eq	67.0	0.595	66	23.1	83.6	96.5
	ax	68.7	0.319	41	14.6	—	—
SF_4	eq	70.7	0.342	94	21.5	84.5	129.7
	ax	72.6	0.268	49	15.6	—	98.6
SOF_4	eq	70.5	0.381	89	21.9	83.0	—
	ax	71.5	0.381	71	19.1	—	—
ClF_5	ax	65.3	0.866	78	29.4	82.8	97.2
	ba	66.2	0.619	64	23.8	—	—

Nonbonded Charge Concentrations:

Molecule		r (pm)	$-\nabla^2\rho$ (au)	Area $(10^{-2}\,pm^2)$	Thickness (μ) (au)	nAn (°)	Net charge on F
ClF_3	eq	60.5	1.58	152	61.9	147.8	−0.30
	ax						−0.52
SF_4	eq	66.6	1.529	190	30.7		−0.71
	ax						−0.74
SOF_4	eq[a]	46.3	1.175	206	30.8		−1.29(O)
	eq						−0.70(F)
	ax						−0.72
ClF_5	ax	60.0	1.900	135	69.9		−0.30
	ba						−0.44

[a]SO bond.

concentrations, while in the plane through the three bonds there are three bonding charge concentrations and a fourth maximum, which represents the saddle between the two nonbonding charge concentrations in the equatorial plane. The size and magnitude of these charge concentrations decrease in the order nonbonding > equatorial bonding > axial bonding. The angle between the two nonbonded charges concentration is 148°, which is fully in agreement with the VSEPR model, which predicts this angle to be larger than the "ideal" angle of 120°. The angle between the bonded charge concentrations is 83.6°, again in agreement with the prediction of the VSEPR model that this angle should be less than 90°.

     In SF_4 there are similarly five charge concentrations in the valence shell of sulfur with the nonbonded charge concentration in the equatorial plane (Figure 7.20). The angle nAb(eq) is 129.7° and the b(eq)Ab(eq) angle is correspondingly less than 120°. The magnitudes and sizes of the charge concentrations decrease in the same order as in ClF_3. There is a similar pattern in SOF_4. That the axial bond charge concentrations in these molecules are farther from the nucleus than the equatorial charge concentrations is consistent with the VSEPR model, which predicts that, because the axial electron pairs are in a more crowded situation than the equatorial pairs, the axial pairs will be pushed farther away from the core. This causes the axial bonds to be longer and more polar than the equatorial bonds. The fact that the axial charge concentrations are smaller in size and

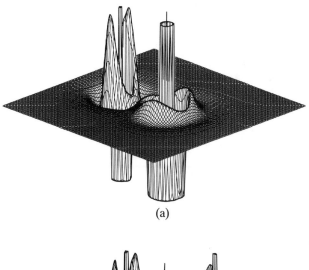

(a)

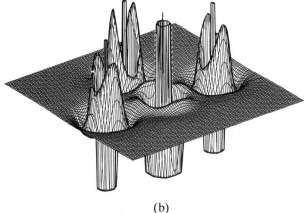

(b)

Figure 7.19 Negative Laplacian of the electron density for ClF_3 (a) in the equatorial plane and (b) in the plane through the three bonds. (Reproduced with permission of R. F. W. Bader)

magnitude than the equatorial charge concentrations is consistent with more charge being transferred to the fluorines in the axial positions. The integrated net charges on the axial fluorines in ClF_3 is -0.52, compared with -0.30 on the equatorial fluorine.

Valence Shells Containing Six Charge Concentrations

Table 7.6 includes data for ClF_5, which has an AX_5E square pyramidal geometry. There are six charge concentrations in the valence shell of the chlorine, five bonding and one nonbonding, with an approximately octahedral arrangement. The nonbonding charge concentration is larger in area and magnitude and closer to the nucleus than the bonding charge concentrations. The four bonding charge concentrations in the base of the square pyramid are farther from the nucleus than the fifth axial charge concentration. This corresponds to the prediction from the VSEPR model that the bonds at the base of the square pyramid are expected to be longer than the single axial bond. Again the charge concentrations

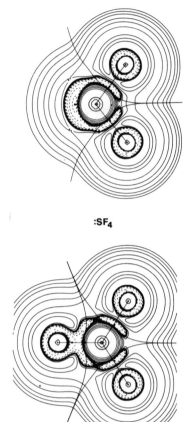

:SF₄

O=SF₄

Figure 7.20 Contour plot of the Laplacian of the charge density for the SF₄ and OSF₄ molecules in the equatorial plane. (Reproduced with permission of R. F. W. Bader)

corresponding to the longer bonds are smaller in area and magnitude than the charge concentration of the shorter bond, and the fluorines in the base of the square pyramid have a larger net charge than the axial fluorine.

Transition Metals

In our discussion of the geometry of transition metal compounds in Chapter 6 we mentioned the cases of $VOCl_3$ and CrO_2Cl_2 and similar compounds in which the bond angles made by the M=O double bonds are smaller than the tetrahedral angle rather than larger as predicted by the VSEPR model and as is observed for compounds of the main-group elements. Calculation of the Laplacian of the electron density for $VOCl_3$ and CrO_2Cl_2 has however shown that this apparently anomalous bond angle can be accounted for in terms of the nonspherical shape of the outer shell of the core ($3s^2 3p^6$ in V^{5+} and Cr^{6+}) caused by its interaction with the bonding pairs in the valence shell. Figure 7.21 shows a section through the negative Laplacian of the electron density for $VOCl_3$. The outer shell of the core

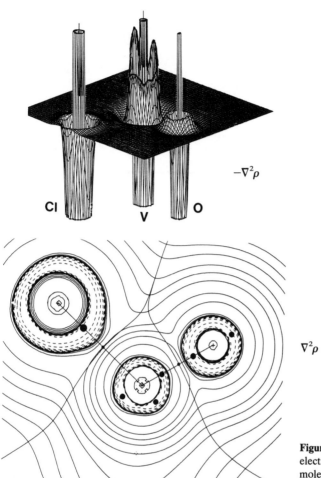

Figure 7.21 The Laplacian of the electron density for the VOCl$_3$ molecule in a VOCl plane. • Maximum of the charge concentration. (Reproduced with permission of P. J. MacDougall)

is not spherical but has a charge concentration opposite each of the bonds. These charge concentrations in the core may be called antibonding charge concentrations. They arise from the repulsion between the bonding charge concentrations and the electrons in the outer shell of the core. The antibonding charge concentration opposite the V=O bond is larger than the three opposite the V—Cl bonds because the V=O bonding charge concentration is larger than the V—Cl bonding charge concentrations. This larger antibonding charge concentration repels the V—Cl bond charge concentrations and opens up the V—Cl angle. Exactly similar arguments apply to the CrO$_2$Cl$_2$ molecule in which the two antibonding charge concentrations opposite the Cr=O bonds are larger than those opposite the Cr—Cl bonds.

In Chapter 6 we accounted for the geometry of molecules of the transition metals in which there is a nonbonding d subshell on the basis that, except for the

d^5 (five unpaired electrons) and d^{10} configurations, the d subshell is nonspherical and has either an oblate or a prolate ellipsoidal shape. In the case of an AX_3 molecule such as VCl_3, we pointed out that such a nonspherical d shell would not be expected to affect the geometry. Thus VCl_3, in which vanadium has a d^2 configuration, would be expected to have a regular equilateral triangular AX_3 geometry. The calculated geometry of VCl_3 is indeed planar with 120° bond angles, and the Laplacian of the electron density (Figure 7.22) shows that there are five charge concentrations in the outer shell of the core, three opposite each of the bonds and one above and one below the plane of the molecule, the five charge concentrations thus having an overall trigonal bipyramidal geometry. Since the axial charge concentrations are larger than the equatorial charge concentrations, the assumed ellipsoidal shape of the core is a reasonable approximation.

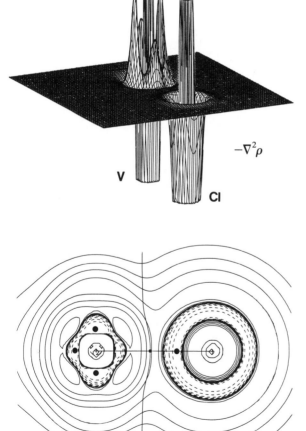

$-\nabla^2\rho$

V

Cl

$\nabla^2\rho$

Figure 7.22 The Laplacian of the electron density for the VCl_3 molecule in a plane through a VCl bond perpendicular to the molecular plane. (Reproduced with permission of P. J. MacDougall)

The core of a transition metal atom in a compound consists not only of a noble gas core but also in most cases of some d electrons. It is convenient to think of these d electrons as constituting a d subshell between the noble gas core and the valence shell. In general, therefore, except for the d^0, d^5 (five unpaired electrons), and d^{10} configurations, the d subshell is nonspherical. Moreover, because it is incomplete it is more polarizable than the $1s^2$ and $2s^2 2p^6$ cores of the lighter main group elements. Thus in most cases the core of a transition metal atom is both nonspherical and more polarizable than the $1s^2$ or $2s^2 2p^6$ cores of the main group elements of periods 2 and 3 and it can, in many transition metal molecules, distort the basic AX_n geometry.

Further studies of the Laplacian of the electron density of transition metal molecules should greatly improve our understanding of their geometry.

A PHYSICAL BASIS FOR THE VSEPR MODEL

Although the total electron density does not provide any evidence for the localized electron pairs postulated in the Lewis and VSEPR models, the number and properties of the local charge concentrations in the valence shell of an atom in a molecule that are revealed by the Laplacian correspond very closely to the number and properties of the localized electron pairs assumed in these models. It is reasonable to assume, therefore, that these charge concentrations provide a physical basis for the Lewis and VSEPR models. The concept of a localized electron pair implies that there exists a region of real space in which there is a high probability of finding two electrons of opposite spin and for which there is a correspondingly small probability of exchange of these electrons with the electrons in other regions. That there is a tendency for electrons to be associated in pairs is a consequence of the Pauli exclusion principle. The usual statement of this principle is that no two electrons in an atom can have the same four quantum numbers, but a more general statement is that the total wave function for a many-electron system must be antisymmetric to electron interchange; in other words, the wave function must change sign when the positions of any two electrons are interchanged. The physical consequence of this restriction on the form of the total wave function is that electrons of the same spin have a high probability of being found apart in space, a low probability of being found close together, and a zero probability of being found at the same point. Thus, if an electron of a given spin α has a high probability of being found at some point in a molecule, there will be a region of space surrounding this point where there will be only a small probability of finding another electron of the same spin; this region of space is called a *Fermi hole*. But there is no restriction against an electron of opposite spin from entering this space. An electron of opposite spin β will similarly exclude other electrons of the same spin as itself, so there is a high probability of finding two electrons of opposite spin in this region and a low probability of finding any other electrons. In other words, there is a high probability that one electron pair will occupy this region of space. The correlation between the distributions of electrons with the same spin is called *spin* or *Fermi correlation.* Spin correlation does not directly cause electrons to pair up. Rather, since there is no spin correlation between an electron of α spin and an electron of β spin, an $\alpha\beta$ pair is obtained as a result of all other electrons being excluded from a region of space in which the $\alpha\beta$ pair is bound by some attractive force. Coulomb repulsion between the electrons, however, limits the localization of the electrons. When the attractive force is large, as for the 1s core region of an atom, the electrons are in a deep potential well and the localization of the electron pair is almost complete. But the Coulomb repulsion between the electrons acts to disrupt this correlation when the attractive force is weak, that is, when the electrons are in a shallow potential well. Electron pairs that are reasonably well localized are found in a few molecules such as LiH, BeH_2, and BH_3 in which, in addition to the pair of $1s^2$ core electrons, one, two, and three pairs of electrons are, respectively, 96%, 93% and 82% localized in the valence shell of each proton.

Although there is localization of electron density into shells in a free atom, as is reflected in the Laplacian of the electron density, there is no possibility of defining localized pairs within a shell because the Fermi hole for each electron is uniformly spread over the respective shell. We have seen that the approach of ligands to a free atom disrupts the uniform charge distribution of the valence shell and causes the formation of local charge concentrations. It is the ligands that provide the attractive force needed to overcome the Coulomb repulsion between the electrons and thus cause the formation of electron pairs. However, the decrease in potential provided by the ligand is in general not sufficient to localize a pair of electrons to the same extent as in the core region of an atom. The tendency toward the formation of localized pairs of electrons on bond formation is illustrated in an approximate manner in Figure 7.23. Here we consider a

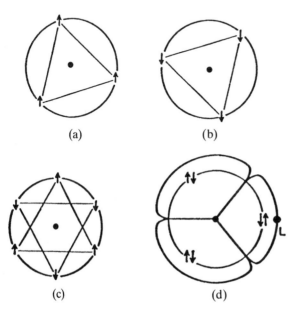

(a)   (b)

(c)   (d)

Figure 7.23 Spin correlation. (a) Set of three electrons with the same spin confined to a circle. (b) Another set of three electrons with opposite spin to those in (a). (c) Spin correlation maintains the most probable triangular arrangement in each set, but the two sets tend to keep apart as a result of electrostatic repulsion. (d) The electrostatic attraction of the nucleus of a ligand L causes two electrons of opposite spin to move toward L and thus to occupy the same region of space so that the electrons become grouped in pairs of opposite spin, one bonding pair and two nonbonding pairs.

hypothetical two-dimensional valence shell in which there are three electrons of α spin and three electrons of β spin. According to the Pauli exclusion principle, the most probable relative arrangement of the three electrons of α spin is when they are as far apart as possible, that is, when they are at the corners of an equilateral triangle. Similarly, the most probable arrangement of the three electrons of β spin is when they are at the corners of another equilateral triangle. In the absence of Coulomb repulsion there would be no correlation between these two sets of electrons, and we can think of each set as free to rotate independently around the nucleus, giving an overall uniform electron density distribution. Coulomb repulsion will, in fact, reinforce the arrangement in each set and will also tend to keep the two sets apart, thus opposing the formation of pairs of opposite spin. The potential provided by a combining ligand will attract one electron of each set into the bonding region to form a bonding pair, and at the same time, as a consequence of spin correlation, two nonbonding pairs are formed. However, the extent to which each pair is localized depends on the strength of the potential field due to the ligand, and it has been shown that in most molecules neither the bonding nor the nonbonding electron pairs are highly localized. In other words, the regions of space where each electron pair is most likely to be found overlap considerably. This is shown, for example, by the Fermi hole of an electron in methane in Figure 7.24. When an electron is close to the nucleus, it has a highly localized Fermi hole, and a second electron of opposite spin can occupy this hole, forming a strongly localized $1s^2$ pair. However, an electron situated at the position of one of the bonding charge concentrations has a much more delocalized Fermi hole, which overlaps extensively with the Fermi hole associated with an electron situated at the position of the bonding charge concentration in an adjacent bond. Moreover, if an electron is close to a line bisecting the angle between two CH bonds, then its Fermi hole extends over both hydrogen nuclei. Thus it is not possible to associate each electron pair with a separate region of space to which it is completely confined.

The Fermi hole for an electron located at the position of a bonding maximum in the VSCC of the carbon atom in methane has the appearance of an sp^3 hybrid orbital or of a localized bonding orbital of molecular-orbital theory. This resemblance of a Fermi hole density to that of a localized valence orbital is obtained only when the reference electron is placed in the neighborhood of a local maximum in the VSCC. The Fermi hole is much more delocalized for other positions of the electron throughout the valence region. Nevertheless, there is partial localization of the valence density in methane and other molecules. The condensation of the electron density into four partially localized pairs of electrons along the four tetrahedral axes of the methane molecule is a result of the combined effects of the ligand field and the Pauli exclusion principle.

Thus, although in most molecules the electrons in the valence shell of an atom are not localized in pairs in separate regions of space, there are local charge concentrations that exhibit all the properties that have been ascribed to localized electron pairs in the VSEPR model. This means that the same arguments that are used in the VSEPR model to predict many features of molecular geometry are still valid in a modified VSEPR model that ascribes these properties not to fully

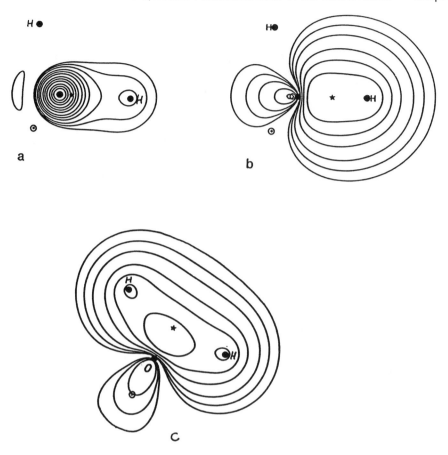

Figure 7.24 Fermi hole of an electron situated at various positions ★ in the methane molecule: (a) close to the carbon nucleus; (b) on the axis of a CH bond close to a maximum in the VSCC; (c) on the bisector of an HCH angle. (Reproduced with permission of R. F. W. Bader)

localized electron pairs but to local charge concentrations. The basic postulate of the VSEPR model that the geometry of a molecule is determined by the arrangement of electron pairs that maximizes their separations has to be replaced by the postulate that the most stable geometry of a molecule is that which maximizes the separations between the local charge concentrations. No rigorous proof has yet been given of this basic postulate, but all other aspects of the VSEPR model are now firmly linked to the properties of the charge density of a molecule. The charge density is a physical property of a molecule that can, at least in principle, be obtained from experimental X-ray crystallographic measurements or, given a sufficiently powerful computer and sufficient time, it can be calculated. The charge density is independent of any orbital model used to describe the bonding in a molecule, and it is in this sense that we can say that the charge density provides a true physical basis for the VSEPR model.

REFERENCES AND SUGGESTED READING

R. F. W. BADER, *Atoms and Molecules: A Quantum Theory*, Oxford University Press, Oxford, England, 1990.

R. F. W. BADER and H. ESSEN, *J. Chem. Phys.*, **80**, 1943, 1984.

R. F. W. BADER, R. J. GILLESPIE, and P. J. MacDOUGALL, *J. Amer. Chem. Soc.*, **110**, 7329, 1988.

R. F. W. BADER, R. J. GILLESPIE, and P. J. MacDOUGALL, in *Molecular Structure and Energetics*, Vol. 11, J. F. LIEBMAN and A. GREENBERG, Eds., VCH, New York, 1989.

R. F. W. BADER, P. J. MacDOUGALL, and C. D. H. LAU, *J. Amer. Chem. Soc.*, **106**, 1594, 1984.

C. A. COULSON, *Valence*, 2nd Ed., Oxford University Press, Oxford, England, 1961.

R. DAUDEL, *The Fundamentals of Theoretical Chemistry*, Pergamon Press, Elmsford, N.Y., 1968.

J. E. HUHEEY, *Inorganic Chemistry*, 3rd Ed., Harper & Row, New York, 1983.

J. N. MURELL, S. F. A. KETTLE, and J. M. TEDDER, *The Chemical Bond*, 2nd Ed., Wiley, Chichester, England, 1985.

A. SCHMIEDEKAMP, D. W. J. CRUICKSHANK, S. SKAARUP, P. PULAY, I. HARGITTAI, and J. E. BOGGS, *J. Amer. Chem. Soc.*, **101**, 2002, 1979.

A. F. WILLIAMS, *A Theoretical Approach to Inorganic Chemistry*, Springer-Verlag, Berlin, 1980.

Subject Index

Formula Index

P

S